160 Fair Years

The Story of the Delaware County Fair
1841 - 2001

160 Fair Years

The Story of the Delaware County Fair

1841 - 2001

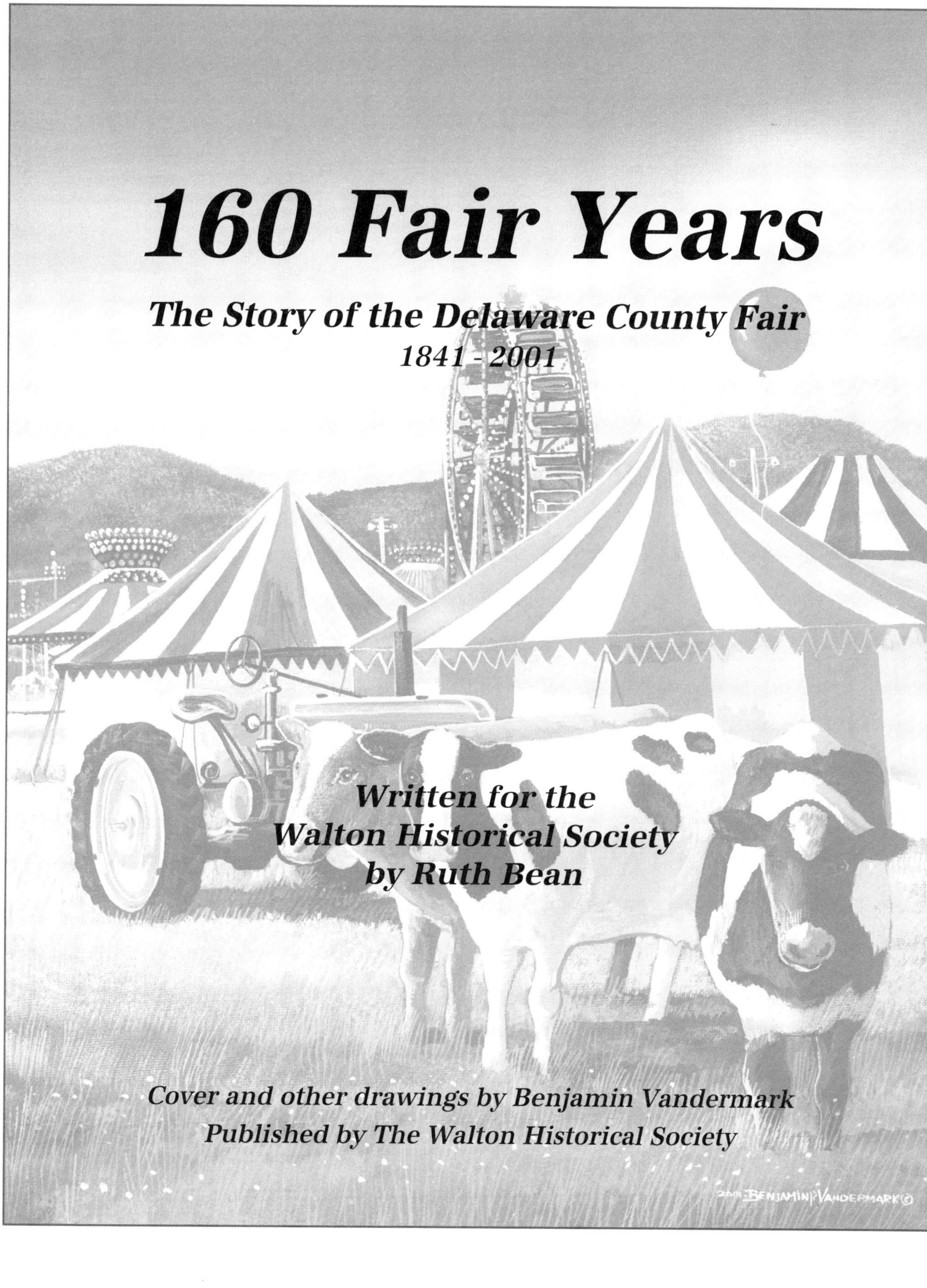

Written for the
Walton Historical Society
by Ruth Bean

Cover and other drawings by Benjamin Vandermark

Published by The Walton Historical Society

First Edition 2001
Printed in the United States of America
By: The Reporter Company, *Printers and Publishers, Inc.*
181 Delaware Street, Walton, New York 13856

ISBN 0-9713118-0-3

Printed in the U.S.A.

Acknowledgments

Until now, no comprehensive history of the Delaware County Fair, which involves both Walton and Delhi, has been compiled. It is to the credit of the Walton Historical Society to suggest that this be done, with a special garland for its historian, Beatrice Bennett. To tell the story would, of course, have been impossible without the help of many people. The long list of those who provided assistance in one way or another would require another volume.

But especial thanks should be given the fair board for its interest and assistance, particularly: Gary Tweedie, Mundale neighbor and long-time friend, without whose enthusiasm and assistance the task could not have gone so smoothly; John Jackson, Fair Board President, whose help in obtaining historical information was invaluable; and Joe Hoyt, Fair Board Secretary, who kindly provided access to materials which clarified events at numerous points in history. Particular thanks to Sam Swart, Walton Historical Society President, for his constant encouragement and to Bert More for a treasure trove of photographs.

I am indebted to Gail Fuller at the New York State Department of Agriculture and Markets for information about certain ancient events and to Walter Rich for his time and the material he provided. Veda Thomson graciously lent cherished pictures and records of by-gone times. Greatly appreciated is the kindly interest of John Hayes; the materials lent by Hugh Sutherland, Linda Smith, Arnold Froehlich and Patty Lollot were invaluable. Thanks go to Anna Sulger and Karen Fletcher at the William B. Ogden Free Library, to personnel at the Cannon Library in Delhi and to Patricia Breakey and Mark Boshnak at The Daily Star, Oneonta, and to Carole Sbordone, Mary Loker Hafele, and Judy Fischer at The Walton Reporter.

Many Waltonians and others helped verify facts and provided pictures, including, but certainly not limited to, Lois Brandow, Kate Henderson, Pam and Greg Graupman, Jeri and Ed Brayman, Richard and Nancy LaTourette, Patsy Moody, Cathy Moore, Arlene Lurenz, Vanley and Alice Blackman, Irma and Leo Pierce, Joyce Butcher Webler, William R. O'Dell, Jr., Helen Schriver Zandt, Jim Northrup, Marge Schreier, and, certainly not least, Jim Rice.

— *Ruth Bean*

This book is dedicated to all who love
the land and its animals and wish to preserve them,
and to a certain farm family,
beloved but gone, from Fish Hollow

Contents

Part I:
The Fair Years

Part II:
Under the Tents and Inside the Track

Appendix

Forward

Imagine long lines of camels laden with silks, spices, foodstuffs and costly fabrics making their way into a center already teeming with turbaned merchants and colorfully dressed buyers hotly bargaining over all kinds of merchandise.

Now cut to the stop-and-go line of traffic on South Street, Walton, during the third week of August any time during the past century and you have come full circle in an ancient human enterprise known as The Fair. The delight of buying and bargaining, showing off livestock and products, competing for prizes and just having a good time with friends had its beginnings in the deep, dark reaches of early society, very likely before the days of the pharaohs and almost certainly before Biblical times.

Fairs were a way to put supply and demand in one place many centuries before fast transportation and efficient marketing became available. In the days of Rome, caravans brought in all kinds of products from the far-flung reaches of the Empire. The Romans introduced fairs into northern Europe in order to encourage trade in the lands they had conquered. After the Western Roman Empire fell in the late 400s, there was a period of two centuries without any commerce to speak of. In the seventh century, however, under Charlemagne, fairs again became important. In particular, people congregated for religious festivals; for several hundred years the fairs in Europe and England were held largely under the sheltering arms of the Church.

The Ellis Peters mystery novel, "St. Peter's Fair" presents a lively picture of a 12th-century fair in Shrewsbury as the detective-monk, Cadfael, goes about dealing with murder. Traders of all kinds came from everywhere, up the river, through the medieval forest, from Ireland, Wales and even France, to offer their goods in Shrewsbury. Into the great court of the abbey came the gentry and yeomen, the rich to buy fine cloth, imported wines, preserved fruits, tapestries and treasures made of gold and silver. Donnybrook was a fair, too, a wild one, which is how the word got into our language!

But as commerce became more organized, the great medieval trade fairs began to die out. The Church stopped promoting fairs in the mid 1500s. Some fairs disappeared because of excessive taxation by greedy rulers; others were abolished. (In Russia all 17,500 fairs were done away with in 1930 because the Soviet Union deemed them "unsuited to the Soviet economic and political objective".) And with the coming of the Industrial Revolution fairs became unnecessary – or their character changed.

As early as the 17th century, though, agricultural and livestock fairs began to arise as a substitute for the trade fairs of medieval times. The Dutch were responsible for the first fair in the American colonies. In 1641 the government of New Netherland authorized an annual fair in New Amsterdam; with the arrival of the English, their tradition of fairs was transported to the colonies. By the mid-1700s fairs had become popular on this side of the Atlantic.

According to records at the New York State Department of Agriculture and Markets, the first State Fair was held in 1811 in Pittsfield, Mass. The first County Fair in New York State took place in 1816 in Cooperstown, Otsego County, and the second in 1817 in Jefferson County. The name of Elkanah Watson is linked inseparably to these events.

Mr. Watson is considered by historians as "the father of the modern agricultural fairs" and his book, *History of the Western Canals and of Modern Agriculture Societies*, published in 1820, is the history of their beginnings. In a 1942 article entitled *Elkanah Watson and New York's First County Fair*, Clifford Lord has written: "Certainly no other institution was nearly so important a vehicle as the Farmer's Holiday in spreading knowledge of the Agriculture Revolution. Certainly no other institution for years could rival the fair as a social and educational influence among the farming population, when this was, above all, a nation of farmers."

Mr. Watson, born in Massachusetts in 1758, moved to Albany after the American Revolution when he was thirty-one, moved to Pittsfield in 1807, and returned to Albany in 1816, working on the forefront of the agricultural movement all this time. His pair of imported merino sheep, tied to an elm tree in the Pittsfield public square, is sometimes cited as "the first American fair in which exhibits were the prime object." He founded an agricultural society in Berkshire County, Mass., which held the 1811 fair. When he returned to Albany, several counties in New York State were hoping to establish agricultural societies and hold fairs, and they asked Watson to help them. His first response was to Otsego County.

Impressed by these developments, Gov. DeWitt Clinton in 1819 recommended the creation of a state agricultural society, a board of agriculture and a pattern farm, and the legislature that year appropriated $10,000 a year for two years, the money to be used by agricultural societies and fairs under the jurisdiction of a Board of Agriculture. Under the influence of Elkanah Watson and the incentive offered by the state funds, six counties formed agricultural societies before the end of 1819. Delaware, which did not form a society at that time, would have received $200.

In 1832 the New York State Agricultural Society was founded at a State Convention held in Albany. An appropriation to support a State Fair was apparently requested, but, again, nothing was forthcoming until 1841 when a new constitution was written. It provided for the distribution of funds to all of the counties "for the promotion of agriculture and households arts in the state" as well as $700 to be used by the state society. With this money, the society held a cattle show and fair in the Village of Syracuse on Sept. 29 and 30, 1841. The State Fair thus was organized and it was held thereafter in various cities:

Albany – 1842, 1846, 1850, 1859, 1871, 1873, 1876, 1880, 1885 and 1889
Rochester – 1843, 1851 (100,000 attended!), 1862, 1864, 1868, 1874, 1877, 1883 and 1887
Poughkeepsie – 1844
Utica – 1845, 1852, 1863, 1865, 1870, 1879, 1882 and 1886
Saratoga – 1847, 1853 and 1866
Buffalo – 1848, 1857 and 1867
New York City – 1854
Elmira – 1855, 1860, 1869, 1872, 1875, 1878, 1881, 1884 and 1888
Watertown – 1856, 1861
Syracuse – 1841, 1849, and 1858

In 1889 Syracuse offered the Society perpetual use of 100 acres of land, and in 1890 that city became the permanent home of the State Fair.

In the United States and Canada today, fairs are a major industry which, according to the World Book Encyclopedia, earn more than $1.7 billion for the 3,200 areas in which they are held. The purpose of this book is to document the course of The Fair in Delaware County through the years with particular emphasis on the history of the Delaware Valley Agricultural Society and the Walton Fairground which has been the site of the county fair for well over a century.

—*Ruth Bean*

It's Fair Time

It's hurry, hurry, hurry!
We feel it everywhere.
For folks of many ages
Are headed to the Fair.

The ring of joyful voices,
The chimes of the carousel;
All blend in happy fusion
To echo out the call.

The time it passes swiftly
There is so much to see.
Each day is full of pleasures
For folks like you and me.

Too soon the Fair is over,
It seems so hard to bear.
Of one thing we are certain;
next year we'll all be there.

by Norma Vigus
1995

Part I

The Fair Years

The First County Fairs

The year 1841 marks the beginning of history for agricultural fairs in Delaware County. On May 5 of that year the New York State Legislature passed an "Act for the Encouragement of Agriculture" with an appropriation of $8,000 to be divided among the various counties. This gave official sanction to a movement that had been growing since Elkanah Watson helped Cooperstown hold its fair in 1816 and Gov. DeWitt Clinton proposed the creation of a state agricultural society three years later.

The New York State Agricultural Society, organized in 1832, had formed a network of county societies of which, in 1841, the Delaware County Agricultural Society in Delhi became one. Its share of the appropriation was $106, and accordingly the first Delaware County Fair was held on October 12, 1841, at the home—or rather inn—of Charles L. Judson at the corner of Main and Meredith Streets, Delhi.

According to *The Delaware Gazette*, one of the earliest newspapers, this fair "far surpassed the most sanguine expectations..." It was the granddaddy of the County Fair now held in Walton, and it was probably the pattern for the many local fairs that were held during the late 19th and early 20th centuries, until the mantle of County Fair passed to Walton in 1926.

To understand what it meant to have a fair in 1841 almost calls for a time-machine trip into another age. Delaware County today is rural; then it was wild. Bears, wolves and panthers roamed the heavily-wooded hills, their hides bringing legal bounties.[1] Tanning and lumbering were peak industries. By 1838 there were sawmills at one-mile intervals all along the West Branch of the Delaware.[2] Crews roamed the heavily-timbered hillsides to clear the virgin hemlock stands, strip the bark and send it off to the tanneries—Tannersville bears witness to this era. At the same time, the tall trees of Delaware County became the masts and superstructures of great wooden ships. The huge logs were rafted down the Delaware—often from Deposit, where they had been "deposited" until high water sent them roiling downstream in a slippery and dangerous journey for the lumberjacks. It was a hurly-burly, roustabout time, and many are the stories of rambunctious raftsmen who balanced their dangerous passage downstream on the logs with riotous festivities on the way home. The early settlers had cleared land for sheep raising and dairying, however, and soon agriculture would become the backbone of the region. (Dairying is said to have had its beginnings in Walton about 1825.)[3]

Settlements were small and far apart in 1841. Walton at the time consisted of "a few houses and business places on Delaware Street and Mount Pleasant while all the rest of the present corporation was laid out in farms...." There were two streets, Delaware Street and Gardiner Place, to the east of which was the Village Common, essentially a cow pasture. What is now the fairground was the farm of David More who had arrived in 1839.[4] The population was less than 300. There was a post office, but mail came, by stagecoach, only three times a week. Almost every family kept sheep, spun yarn, and made its own clothing; people read by candlelight, kerosene lamps or wood fires, grew their own products and either killed their own barnyard animals for meat or trapped, fished or hunted the elk, deer, partridge and grouse which were still plentiful. Many families spent the spring and summer making butter which they put into tubs and firkins, loaded on heavy lumber wagons and took to Kingston. Naturally, there was no such thing as just picking up the telephone. And then there were the roads!

Roads in Delaware County, until fairly recent times, have had a dreadful reputation. About 1843 a correspondent to the *Gazette* wrote that "the horrible roads" were "sufficient to deter one from settling in the county." In 1839 a new line of stagecoaches had begun a run from Delhi to Kingston three times a week; the trip took a whole day, 74 miles in 16 hours. The Erie Railroad had been incorporated in 1832 but construction was delayed until 1848 because of the difficult terrain.[5] It took a considerable amount of optimism to plan a fair in 1841 because it would have been a major feat for exhibitors and spectators alike to get there. As a matter of fact, a good number of them didn't! In its 1847 report to the New York State Agricultural Society, the Delhi group complained that "the most populous and distant towns have failed in most cases to be represented."

Finally, there was the economic and political climate to be reckoned with. There had been two years of depression, in 1835 and 1836, although better times had now begun. However, 1839 had seen the beginning of the famous Anti-Rent troubles which were to last, off and on, for a decade as farmers on homesteads owned by distant landlords resisted the feudal system of paying rent to a lord of the manor and demanded the right to own the land themselves. The constitutional amendment that changed the system was not enacted until 1849.

So, an agricultural fair in 1841 would have had to deal with all of these things: A still wild and woolly environment, an unstable political situation, a small population just beginning to feel sure about its agricultural possibilities, difficult transportation, and a still primitive communications system.

On the other hand, many in the 1840s regarded agriculture with a passion approaching religious zeal. In its annual report for 1847, the New York State Agricultural Society president, George Vail, remarked upon the organization's determination to "urge forward, by every wise and thorough effort, the advance of agriculture as the greatest good it can render to the greatest number." And the much-respected ex-Governor Silas Wright wrote: "Agricultural production is the substratum of the whole superstructure...agricultural production is the first in order, the strongest in necessity, and the highest in usefulness, in (the) whole system of acquisition."

With enormous enthusiasm the state society had set about making agricultural surveys of all the counties, and in its fat books of Annual Transactions are notes from farmers describing their experiences with soil conditions, drainage, diseases and insect pests; root crops and corn, oats, barley, beans and flax; the "manufacture" of butter and cheese, the cultivation of fruit trees, the trials of sheep and dairy farmers (130,000 sheep in Wayne County, but wool was worth only 28 cents per pound and milk went for 1.75 cents to 3 cents a quart); growing efforts to improve breeds of cattle and horses, and plans for buildings such as a farmhouse or piggery with detailed illustrations. Fertile lands in many parts of New York State were opening to farming; it was going to be the focus of an era.

It was at this time, also, that the idea of a state agricultural school began to surface, "an institution where scientific knowledge and practical skill might blend their powers in elevating the character of our farming and our farmers, to the high standard which they ought and are able to attain." Enthusiasm, optimism, confidence, a state of mind like that of a prospector viewing an inexhaustible vein of gold, were characteristic of the period.

So the farmers centered in Delhi held a fair. It began with a meeting of a group of prominent citizens in the Court House on July 15, 1841. Charles Hathaway of Delhi was the presiding officer, and Platt Townsend ("Old Doc Townsend," one of the founders of the Village of Walton) was the secretary. The new Delaware County Agricultural Society's charter listed 35 signers. The original charter officers were:

Samuel A. Law of Meredith, president. His was one of the earliest families to settle in the Delhi area.

Daniel McFarland of Delhi, secretary and Platt Townsend of Walton, treasurer.

The vice presidents were Levi Hanford, Jr., Walton; George Sturges, Delhi; Samuel Andrews, Hamden; Thomas Hymers, Meredith; John Raitt, Bovina; and George W. Downs, Colchester.

Some weeks later the group met again and set the date. A. M. Paine, editor of the *Gazette*, reported that "provision will be made for the officers and members of the society to dine together," and premiums were to be offered "in the amount of $3, $2, and $1 on Neat–Cattle, Sheep, Swine, Horses, Maple Sugar, Crops."

After the event was over, the *Gazette* announced that the fair had been "numerously attended," and Mr. McFarland, the secretary, expressed his belief that it had "accomplished two very important objects: It has excited the attention of our farmers and it has clearly demonstrated the practicability of Agricultural Improvement. These are two of the grand objects for which such associations are organized. Delaware County is behind the age in improvement, but it required just such an impulse to place her at least on an equality with her sisters." An added note: "Specimens of Maple Sugar...were remarkably fine. First premium went to Isaac Burr of Meredith."

When the Agricultural Society held its first annual meeting in 1842, Jabez Bostwick of Hamden was elected president to succeed Samuel A. Law; Daniel McFarland was reelected secretary and George Sturges of Delhi succeeded Platt Townsend as treasurer.

In its 1847 report to the New York State Agricultural Society, the Delhi group called that fair "one of the best held in the county," but the writer added: "There was not so large a display of animals, as on other occasions, from the fact that many of our best towns have not been represented this year....The annual fairs have, by consent, been held at Delhi, therefore the most populous and distant towns have failed in most cases to be represented.

"In many of the towns the most worthy and enterprising farmers, mechanics, and merchants, have devoted their time to the manufacture and purchasing of *Lumber*. Many of our hills and valleys in their rude and original state, were clothed with the pine, hemlock together with the cherry, hard maple, and various other hard woods, which have nearly disappeared, were taken from our county, and rafted and floated down the Delaware to Philadelphia... There is at this time an interest felt, which if properly directed, will place Delaware among the foremost agricultural counties in the state.

"Her thousand hills, which are enriched by her numberless springs oozing from the very tops, are capable, if properly tilled, of producing not only grass, but grain in abundance...No county can raise better stock or keep more cattle on a given number of acres than Delaware, nor produce more pounds of butter from a single cow."

The 1847 Fair again was held at Judson's Inn "where a spacious room had been provided, which was devoted to showing articles of domestic manufacture and horticultural products," including a "fine display of a cage of 30 canary birds raised by Mrs. P. P. Wright." Delhi had received $106 from the state society, having raised an equal amount itself. It awarded $161.75 in premiums.

In 1851, ten years after it had begun, the Delaware County Fair had grown to a two-day event and its entries had expanded to:

"Improved Cattle, Grade Cattle, Native Cattle, Oxen and Steers; Maple Sugar, Flour and Bread, Fruits and Vegetables, Mechanical Work (offered by the maker), Farm Wagon, Buggy, Harness, Ox Yolk (*sic*), Ox Chain and Set Shoes, Plow for Sod, Fanning Mill, Grain Cradle, Firkin and Churn."

After 1851 the County Fair was held in a variety of locations, the town offering the best inducement usually given the privilege of holding it. Thus, the Fair was held

In 1855, in Hobart
In 1856, in Walton
In 1857, in Andes
In 1858, in Franklin
In 1859, in Bloomville
In 1860 and 1861, in Hobart
In 1862, in Franklin
In 1863, in Delhi
In 1864, in Hobart
In 1865, in Walton
In 1866, in Bloomville
From 1867 to 1871, in Walton
From 1872 to 1879, in Delhi
From 1880 to 1925 in Delhi with The Great Walton Fair occurring simultaneously beginning in 1887.

The County Fair of 1856 in Walton really excited the editors of *The Walton Blade*, that village's first newspaper—*published every Tuesday by Berray & Pine, subscription 75 cents a year, offering correspondence, "Humorous and Sentimental, solicited from all parts of the world."*

The Blade devoted more than generous space to plans for the Fair which would take place on Wednesday and Thursday, October

24 and 25. As early as June it began to run lists of the premiums to be offered. For:

Durham, Devon, Ayrshire, Hereford and "native" bulls, cows and heifers; oxen and steers, sheep, horses, swine and poultry.

Sugar and flour, fruit and vegetables, leather and cloth, farming utensils such as milk pails, harness, rakes and buggies.

Best power for dog or sheep, best farm wagon and power for churning.

Plowing, grain and seeds, winter fruit, dressed meats and poultry, field crops and Ladies' Handwork. There was a class for Ladies' Equestrianship.

The plowing was to be done "without the aid of a driver." Each plowman was required to "plow a land separate from his neighbor and show two back furrows and finish with a dead furrow between."

The premiums ran anywhere from fifty cents to five dollars.

Some familiar family names appear among the list of premium winners: Hawley, Bassett, Hoyt, Miller, Van Volkenburg, Wakeman, Kilpatrick, Hanford, McGibbon, Griswold, Pine, Benedict, Fancher, Tweedey (*sic*), Barlow, Wood, North, Eells, Townsend, Frisbie....

The Fair took place on what was then called the "general training ground," an area near the present location of the Walton Grange Hall and surrounding buildings on Stockton Avenue.[6] It was financed largely by dollar memberships which were collected beforehand. Each member received tickets for his own family; others paid twelve and a half cents admission. Should the Fair receipts not be equal to the amount of premiums to be awarded, the Delaware County Agricultural Society ruled that the premiums would be reduced across the board.

Richard B. Gibbs of Harpersfield was president of the Society for the 1856 Fair, and vice presidents were George W. Hanford, Walton; Russell D. Baird, Harpersfield; Porter Frisbie, Meredith; Samuel F. Miller, Franklin; Daniel Andrus, Stamford; Ira E. Sherman, Sidney; and Henry C. Hawley, Hamden. The Walton judges included John Mead, S. H. White, William A. White, Frederick Eells, Thadeous Hoyt and Amasa Hoyt.

The Judges on Animals were instructed to "have regard for the symmetry, early maturity, size and general characteristics of the animals they examine, making proper allowance for age, feeding and other circumstances and are expressly required not to give encouragement to over-fed animals. No premiums are to be awarded to bulls, cows or heifers which shall appear to have been fattened for the butcher."

After the Fair, a *Blade* columnist wrote: "The Fair is the all-absorbing topic of conversation. Even politics wanes and grows cold before the great exhibition of Delaware County.... Thursday morning every avenue leading to our village was thronged with carriages well filled with human beings. Delaware Street was literally packed for hours with people wending their way to the fairgrounds. The stores were closed, the streets almost deserted and the fairgrounds, enclosing almost five acres, was filled with people and the various objects offered for exhibition. Viewed from any elevated position outside the fairgrounds at this time, the scene was beautiful. The cattle from a thousand hills were gathered within the enclosure, forming a border to the picture; the eight thousand people encircling the ring and filling the carriages and stagings of the great tent; the gay collars of the ladies' dresses fluttering in the breeze, the judges standing on the centre platform with beating hearts waiting to award the premiums, the band sending forth sweet strains of melody, bringing the whole scene into harmony, altogether forming a panorama never to be forgotten."

The Blade reported attendance as "variously estimated from seven to ten thousand" with the exhibition of cattle "never better."

Even in these early years, however, complete harmony was not to be found—either among the communities which vied for the privilege of holding the County Fair or among those who were exhibiting at the fairs and hoping to receive premiums. The weaknesses of human nature were all too apparent. In an advance article for the 1857 Fair, held in Andes, *The Walton Journal*, successor to *The Blade*, suggested that a person's name should not appear on his entry. "The judges should decide upon the articles without the assistance of the names of competitors. In not a few cases it is the name, aided by a little log rolling, which takes the premium instead of the article exhibited.... A. will get a premium on his horse over B., not because his horse is the best, but A. is known; he is somebody, and B. is obscure.... This illustrates a general fact which was fully apparent last year."

In the larger scene, over a period of time, charges of favoritism and unfairness on the part of the Delaware County Agricultural Society caught up with that worthy body, and on March 8, 1862, a new and competing association was born—the Delhi Agricultural and Mechanics' Society.[7] That year it conducted a fair in Delhi even though the County Fair was held in Franklin. In 1863 there were two fairs in Delhi, one by the Delaware County Agricultural Society and the other by the Mechanics' Society. The latter existed for a decade, holding fairs along with the "mother" group. Munsell's History of Delaware County describes it this way:

It (the Mechanics' Society) *had its origin in the alleged unfairness on the part of the County Agricultural Society toward Delhi and some of the towns lying contiguous thereto in the selection of its officers, in the appointment of judges on articles exhibited at its fairs, in the award of premiums, and in other evidences of prejudice and ill-feeling toward the towns alluded to.*

Its first officers were: President, Edmund Rose; Vice President, P.H. Beardsley; Secre-

tary, Norwood Bowne; Treasurer, Anthony M. Paine. Directors were D. G. Landon, L. C. Hollister, Alexander Marble, A. Cook Edgerton and C. Allen Frost.

The Society at once took high rank with the people, soon becoming exceedingly popular, particularly with the farming and mechanical interests. It held frequent public meetings for the discussion of pertinent subjects; and its fairs were well sustained and rendered useful and interesting through the generous display of stock, vegetables, fruits, agricultural implements, specimens of mechanism and art, and ladies' handiwork from all except the more remote towns in the county.

Soon after it was created, the Mechanics' Society bought from Samuel Sherwood a five-and-three-quarter-acre lot in the village of Delhi and "built an agricultural hall, graded a quarter-mile track and made other improvements." Later the Society leased an adjoining piece of land and doubled the size of the track. But after ten years, in the spring of 1872, "it was deemed better for the public interests that this society should be merged in the County Agricultural Society, and negotiations looking to that end were successful, an amicable arrangement being perfected by a lease to the county society's grounds for a term of ninety-nine years (until 1971), with conditions that were entirely equitable and satisfactory."

In September, 1865, a few months after the Civil War had ended, the Delaware County Agricultural Society of Delhi again held a fair in Walton. The weather was "all that could be desired," and "superior music" was furnished by the regimental band of the 144th Regiment. The speakers were the Hon. Ira R. Sherman and the Hon. William B. Ogden, who had moved from Walton and had become the Mayor of Chicago. In its report to the New York State Agricultural Society, the fair organizers wrote:

"The 25th annual fair of this society was held in the village of Walton the last week in September. The grounds, comprising about seven acres, were fitted up with ample buildings, stalls and pens, and a quarter of a mile track, for the accommodation of the exhibitors and visitors present. The exhibitors were numerous and the competition very strong in nearly all of the departments for the premiums offered.

"In the horse department there were 65 animals entered. The show of matched horses included 14 span and nearly all worthy of a premium. The exhibition of horses for numbers, quality, style and action never has been equaled in the county.

"There were 155 head of cattle presented. The Durham, Devon and Ayrshire were well represented, but the competition among the grades was much the strongest, as they were quite numerous.

"The show of working cattle and steers, contrary to the expectation of all, was extensive and superior... The cattle were brought within the track and exhibited and received their full share of attention.

"A spirited drawing match, conducted upon the principles of humanity, came off the second day and was so arranged as to be witnessed by all who wished to see.

"The season had been favorable for the maturity of vegetables, and this department was unusually attractive, Vegetable Hall, a building 100 feet long, was well filled with an extensive variety. In addition to the usual premiums, the society had offered liberal premiums which would have been a credit to the society, exclusive of all in the vegetable hall.

"The boys' department was well sustained. The ladies competing and the number of articles presented by them was larger than usual and was an interesting part of the exhibition."

There is also a record of the 1869 county fair held in Walton from September 14 to 16. According to the July 3, 1869 issue of *The Delaware Republican* of Delhi, John B. Yendes of Delhi was president of the county society at that time.

The vice presidents were T.S. Hoyt of West Brook, E. Van Dyke, Delhi; E. M. Smith, Meredith; William Miller, Bovina; William Gregory, Sidney; Col. G. D. Wheeler, Deposit and J. A. Bowers, Franklin. G. W. Hanford of Deposit was treasurer and Porter Frisbee of Delhi, secretary. Miss Mary Hanford, Deposit, was assistant secretary of the Ladies' Department and Col. Wheeler was the general superintendent.

Cattle were under the jurisdiction of T. S. Hoyt of West Brook and J. D. Rogers of Walton

Sheep, Alexander Tweedie, Walton.

Swine and Poultry, Stephen Berray, Jr., Walton.

Horses, George St. John, Walton.

Ladies' Department, G. A. Colton and George Parsons, Walton.

Boys' Department, A. J. Hoyt, Walton.

Sugar, Fruit, Flour, etc, J. Cease, Bloomville.

Leather, Cloth and Farming Utensils, John Love, Walton.

There were classes for Durham, Devon, Ayrshire, Hereford and Alderney cattle; Cheviot, Merino, Leicester, Saxon, South Down and Oxfordshire sheep; Essex and Suffolk swine. All had to be "thoroughbred," and they had to have their pedigree with them.

For the butter and cheese classes, the cow had to be "kept on grass only during the experiment and for ten days previous" and contestants had to submit a statement regarding the cow involved—age, time of calving, quantity of milk and quantity of butter made during the trial—and all of this had to be "verified by affidavit."

The following January the Society was to hold its annual meeting, and at that time competitors in field crops would turn in their samples. Required with these would be sworn statements indicating the manner of soil preparation, quantity and quality of manure used, number of bushels of seed used and the weight of the crop gathered.

There was an infinite variety of household products which could be entered, including hats, bedspreads, quilts, yard goods, embroidery, crocheting, knitting, tatting, and oil paintings, "pastile painting," crayon, pencil or pen drawings. Cookery, breads and "assorted baking" also comprised a class.

Mark Boshnack, who has done considerable research into the history of the Delhi fair, reports that in 1879 two fairs were held in Delhi: The Delaware County Fair took place September 16 - 18 and was said to have had "the best exhibit of stock ever known..." The Delaware County Dairymen"s Association also held a fair—their first—on November 12 and 13. There, "100 packages of butter were entered for the various premiums."

An account of the 50th County Fair, held in Delhi in 1890, describes it as having grown to three days. There were grandstand attractions and side shows, and the carnivals "were the finest to be procured through the country." They, of course, were highly featured for they paid steep concession fees. And it was a keen period for horse racing.

A newspaper account declared: "Exhibits of all types were on display and larger premiums were offered to all exhibitors. Cattle groups and cattle individuals brought $10 for first placings. These prices were a great inducement to exhibitors and required buildings to be erected to care for cattle, swine and poultry.

"At the close of the Fair in 1890 the officers were very sure that it had been a definite step in the right direction when they had formed their society in 1841 to give farmers and their families an opportunity to display their yearly efforts and goods for the well-being of mankind, to contest each year for an incentive for families to raise their standards of living and vie in a friendly manner with their neighbors."

Of course the Delaware Valley Agricultural Society in Walton had arisen in 1887, a competing organization. For more than three decades annual fairs were held in both communities, often on consecutive weeks. During this period newspaper ads and printed posters referred to *The Great Walton Fair*, and that name continued to be used long after Walton became the only county fair site.

Delhi has been justifiably proud of its fairground, particularly its grandstand, and over the years a great deal of loving care went into its upkeep and preservation. A local builder, Thomas J. Jackson, was commissioned in 1894 to build it. This grand old structure, according to Delhi historian, John D. Raitt, in his series of books entitled *Ruts in the Road*, was 133 feet long. "The upper part commanded a clear view of the entire grounds while the lower part was fitted for a restaurant as well as for exhibition booths." In 1899 a record crowd filled the grandstand to see and hear Theodore Roosevelt, the future president, who was then governor of New York State. Another gala occasion was the coming, in 1907, of another governor, Charles Evans Hughes, who soon became Chief Justice of the United States Supreme Court. Around 1898 fair officials began to seek a young couple to be married on the grandstand during Fair Week. They finally found one when Sarah Victor and William D. Winters, both of Bainbridge, were married there. In 1904 more than 100 Italian workmen lived in the area under the grandstand for nearly two months while they worked on the construction of Route 10 from the Delhi town line to the mouth of Elk Creek. Later large crowds attended the Fourth of July American Legion celebrations which included high-wire acts and exhibitions by the New York State Trooper Rough Riders.

In 1929 the Delhi fairground was purchased by the Donald W. Gleason American Legion Post #190, which still owns it. Some time during 1957, when it looked as if New York City would acquire the fairground in Walton, the American Legion offered the use of the Delhi site, but, as it turned out, to accept this offer was not necessary.

The last major use of the grandstand was for the Delhi Sesquicentennial celebration in 1971. Subsequently, the legendary old structure was torn down.

Today, the fairground still serves community needs. Turn off Main Street at Wood's Grocery Store onto Gallant Avenue, with the town highway facilities on the right, turn left, and facing you will be a wide, grassy area with a portion of the old race track still visible. The Donald W. Gleason Post is using the ancient but sturdy, white-painted, well-preserved Ladies' Building as its headquarters, and at the north end is a large, well-equipped playground for children.

When there's not a ball game going on, the scene on a summer day, with tall trees on either side of the main expanse, is one of green-grass, cricket-chirping calm and an imaginative wayfarer easily sees the tents and the livestock, the barkers and the running horses and feels the magnitude of things as they were.

Photo courtesy of Hanford Mills Museum, East Meredith, New York

Settlers in the Western Catskills early in the 19th century were pioneers in the most real sense of the word. They built log cabins like this, cooked on open fires and hauled their water from sometimes distant streams. Families raised pigs and chickens for meat, cows for milk and sheep for wool.

Walton Historical Society Photo

Settlements were far apart in the 1840s and the roads were terrible.

Walton Historical Society Photo

This house was so isolated that the family had to take a rowboat and cross the river each time anyone needed to go to school or church, or down to the nearest village to get supplies.

Walton Historical Society Photo

The Walton homestead, built by Joshua Pine in 1804, is typical of housing in the early days of Delaware County.

Photo courtesy of Hanford Mills Museum, East Meredith, New York

A Delaware County winter in the early days was bleak, long and lonely.

Walton Historical Society Photo

Covered bridges, symbol of a slower and more primitive era, crossed the Delaware at many points. Steep hills and poor roads made the county inaccessible for a long time.

FARM HOUSE.

The nearly 800-page bound book, *Transactions of the New York State Agricultural Society for 1847* contains sketches and floor plans of various farm buildings including this farm house. The wing at right includes an ice house, wood room, wash room and "dairy table". The house has a parlor, library, bedroom, dining room and kitchen downstairs and four bedrooms upstairs.

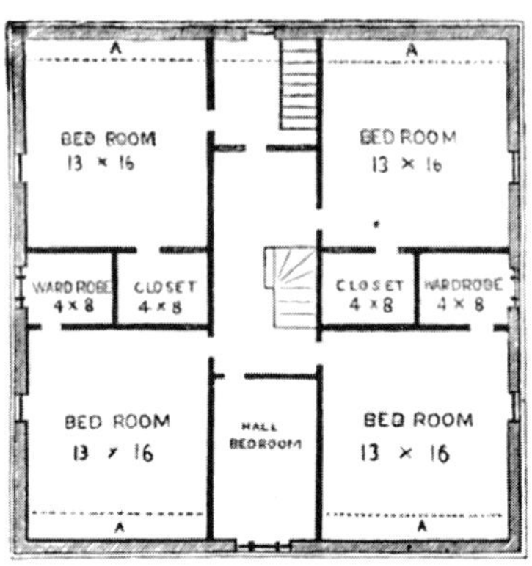

SECOND FLOOR.

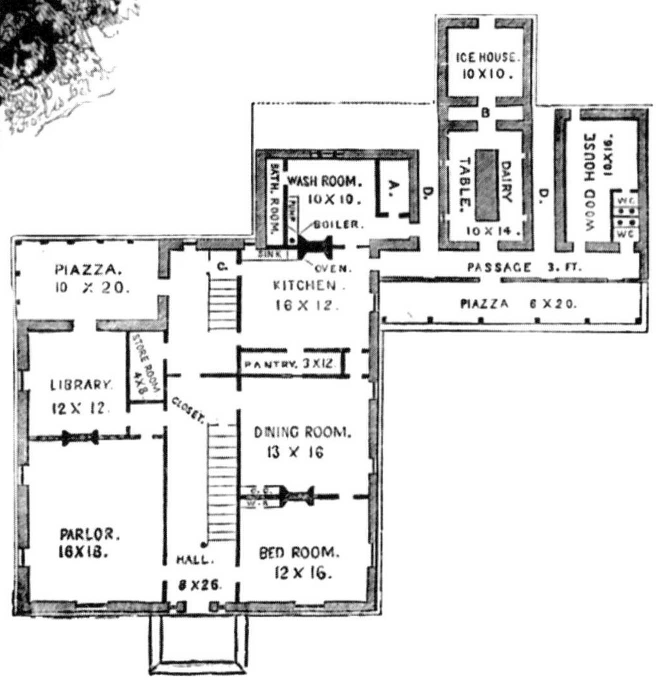

GROUND PLAN.

Photo courtesy of Hanford Mills Museum, East Meredith, New York

In the early days lumbering was a major industry. Masts on some of the famous sailing vessels came from the tall pines cut on Delaware County hillsides and floated down the Delaware. The horse in this picture is working a "horse treadmill" attached to a saw which cuts the logs.

Photo courtesy of Hanford Mills Museum, East Meredith, New York

One big log was enough for two horses to pull to the Hanford Mills.

Walton Historical Society Photo

The rafts which floated huge hardwood logs down the Delaware during the 1800s were a tremendous size. A single raft could be 30 by 160 feet. The raftsmen, who had a reputation for drinking and carousing on their long journey back home, often became legendary local figures.

Walton Historical Society Photo

Eleven men are making the dangerous journey on this Delaware River raft sometime in the early 1800s. The village of Deposit got its name because lumber was often "deposited" there until high water could send the logs, lashed together, down the river to Trenton or Philadelphia.

Walton Historical Society Photo

The Risley lumber sawmill at Rock Rift, Walton, operated for many years. Notice the "clear-cutting" of timber which has been done on the hillside.

A cast iron grain mill of this kind was used in the 1840s for grinding rice, corn, barley, oats and other grains. It was also used on the farm for grinding "hommony" and grain for chickens. It was described as "strong, durable, of simple construction, easily managed and not liable to get out of order." Price, $7 with frame.

–from the *Transactions of the New York State Agricultural Society, 1847*

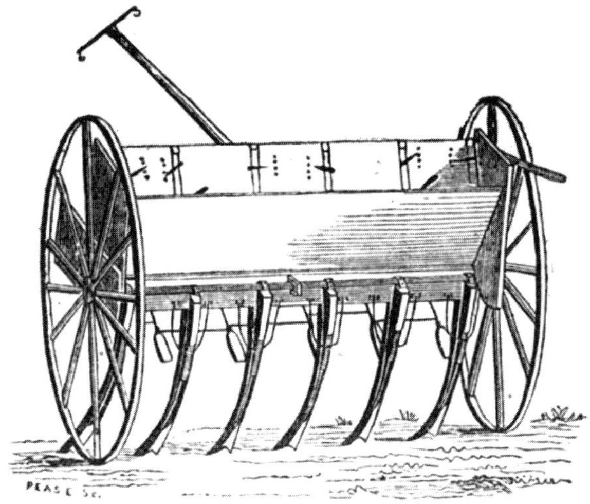

Drawing of a patent lever drill introduced by H. W. Smith of Pennsylvania in 1846. Smith simplified the cumbersome and complicated machines of the time by designing lighter and more adaptable wheels and for the first time providing a graduated index which could be set to pass any given quantity of seed per acre or altered quickly from one quantity to another. It cost $75.

–from the *Transactions of the New York State Agricultural Society, 1847*

Walton Historical Society Photo

A wagon pulled by two sturdy horses was the only way to get the feed sacks from here to there.

–from the *Transactions of the New York State Agricultural Society, 1847*

Today they would be called Shorthorns, but in 1847 they were Durhams. The cow, Hilfa, on the right won first premium at the New York State Agricultural Society fair at Saratoga in 1847. She was owned by George Vail of Troy, president of the Society, and had been sent to him from England. Vail's premium bull, Meteor, is at left with their progeny, a bull calf, in the foreground.

–from the *Transactions of the New York State Agricultural Society, 1847*

Compare these 1847 five-year-old twin steers with animals you might find at the Fair today.

Photo courtesy of Thelma Schriver

Fruits and vegetables were importantly displayed at all of the early fairs.

Walton Historical Society Photo

Looking down from Pine Hill sometime after 1846. The area around Townsend Street is completely undeveloped. The old covered bridge and barns and farm buildings are visible.

Walton Historical Society Photo

When the County Fair of 1856 was held in Walton, many of the visitors and exhibitors probably stayed at Smith's Tavern, corner of Delaware and Bridge Streets. The building is still standing. It was moved to Mead Street, at the foot of Shepard Street and made into an apartment house.

Walton Historical Society Photo

A view of Walton, probably in the early 1870s. The church with the spire is the First United Methodist Church, then called the Methodist Episcopal Church, on North Street. It was built in 1869. The wooden building that was then the United Presbyterian Church, built in 1866, can be seen on East Street at the site of the present parsonage. What are now Bruce and Griswold Streets were still meadow and pasture, and there aren't very many houses on North Street, either.

The Fair in Walton

Photo courtesy of Helen Zandt

"The fairgrounds have been surveyed and a one-half mile track staked out." With these words the Walton *Chronicle* reported the birth of an event that was to give Walton its greatest claim to fame during the next century: THE FAIR!

Toward the end of 1886 a group of farmers and others interested in having a permanent annual fair in Walton had met and begun to make plans. The project stood still for a while because of a pneumonia-type disease which was infecting cattle, but by late spring it seemed possible to go ahead.

On Saturday, June 11, 1887, the group held a formal meeting and discussed the possibility of forming a *Delaware Valley Agricultural Society*. Its name was an echo of its Delhi cousin, the Delaware County Agricultural Society, sponsor of the County Fair since 1841 and destined to continue in that capacity for some years yet. But as the Walton farmers met on that June afternoon, the future belonged to Walton.

The group that met consisted of Charles B. Bassett, cashier at what was then the Delaware County Bank; Joseph Harby, Jr., William C. Doig, William C. Rogers, John H. Townsend, Henry S. White, Robert Harby, Alexander Tweedie, Edwin R. Howland, William A. White and Henry Inderlied.

Before anything could happen, these founders had to file an application with the New York

State Department of State to form a New York State corporation. They created a nine-member Board of Managers consisting of Mr. Bassett, president; Joseph Harby, Jr., Mr. Doig, and Mr. Rogers, vice presidents; Mr. Townsend, treasurer; Mr. White, secretary, and Messrs. Tweedie, Howland and Robert Harby, managers.

On the 29th of June this board signed its request indicating a desire to form "a Town Agricultural Society in the said town of Walton, Delaware County, New York," and "certifying that the name and title of said Society shall be known in Law as *The Delaware Valley Agricultural Society of Walton, New York.*"

The signers stated further that *The particular business and object of such society is for the promotion and improvement of agricultural interests in said County of Delaware and for the purpose of holding Agricultural Fairs in said town.*

According to records at the state Department of Agriculture and Markets, the not-for-profit corporation was legally created on July 12, 1887. In 2001 it is holding its 115th Fair. (Count 1887 as Number One.)

There had been fairs in Walton earlier than 1887. Along with other villages which undertook to hold county fairs under the auspices of the county society, Walton had been the host in 1856 and 1865 and again from 1867 to 1871 when the county fair was held in Walton under the auspices of the County Agricultural Society and in Delhi by the Delhi Agricultural and Mechanics Society.

The fairs in Walton had taken place in a large space along Stockton Avenue and William Street where the Walton Grange Hall and adjacent buildings are now.[1] However, the 1887 organizers wanted something that was larger and could be expanded.

From the Harby Brothers, Joseph, Jr., and Robert, who were "dealers in Live Stock, Hides, Pelts, Wool, Tallow, etc.," the fair organizers leased, at $100 a year, 16 acres that, in the 1830s and 40s, had been the David More farm. They purchased the land in 1897, another small parcel in 1906 from Anna Bell and the last 25 acres in 1968 from Merritt Loudon.[2]

Newspaper articles of the time reveal that one difficulty in promoting the first town fair was the location and condition of the bridge over the Delaware River. The covered bridge, built in 1846, was picturesque but beginning to be inadequate, and townspeople were calling for a new iron bridge to replace it. There was also a proposal to move the bridge from Bridge Street to a point which would constitute a continuation of Townsend Street. That idea never materialized, but on March 30 a contract for $9,000 was let to the Groton Bridge Company for the new bridge. The firm had already built the bridge on Delaware Street, crossing West Brook, as well as a bridge at Delhi. The 200-foot, one-span bridge included a 40-foot stone culvert on the Stockton Avenue side. Work started in July with a September 15 completion date so that fairgoers in October could be sure of a new river crossing. Unfortunately, the bridge did not make it in time for the Fair.[3]

The Fair was scheduled to be held for three days, Tuesday, Wednesday and Thursday, October 4, 5, and 6. Richard Garrison was asked to be general superintendent, and there were 16 superintendents of departments.

According to the July 22, 1887 issue of the *Delaware Dairyman and Register* of Franklin, a committee was appointed to inspect buildings at the Oneonta fairground in Morris "for the purpose of getting information and suggestions to aid in erecting buildings" in Walton.

There were several pleas for funds. The *Chronicle* in July pointed out that "while considerable money has already been subscribed, there is need for those who have not already contributed to do so and that, liberally, in order to make the enterprise a success, as it will take considerable money to erect the necessary buildings, enclosures for stock, Etc." In August, an editorial reminded readers that "everyone should remember that if sufficient funds are raised the grounds will be fitted up complete this year, but if not some of the buildings necessary to the success of a fair will not be built. Let those who have not given bear this in mind."

In any event, the contract for putting up a number of buildings, sheds and fences went to a Robert Murray, and soon workmen were preparing a track also.

The 1887 Fair was an image of its times. In an agricultural countryside, where people lived far apart, where going ten miles to town with a five-mile-an-hour horse was a festive event, and where hay, horses, cows and canning were life's chief interests, a Fair held an attraction virtually impossible for us to realize. There were no professional entertainers, no spiffy horse show, no "rides," no "concessions," no tractor pulls and certainly no demolition derby!

Bert McClenon, who attended that first Fair when he was eight years old—and went to every one after that until he was way past eighty—a few years before his death told a reporter that for many years he went to the Fair from his farm on West Brook by horse and wagon.[4] "There were no good roads," he said. "No electric lights, no telephones." The cattle and horses were not trucked onto the fairground as they are today; instead, it was normal to see a herd of cows being driven through the village on their way to the Fair. In fact, years later, Peter Chambers, whose prize-winning Holsteins were housed on his West Brook farm, would drive his cattle to the Fair.

"There were a lot of ox teams at the early fairs," Bert reported. "My dad had a pair with great big horns." There weren't all these stage acts, he said, but there were the balloon ascensions which gave the crowd a big thrill. And after the fairs got going, there were carnival barkers like Cheap John from Trout Creek who came with a bundle of shoelaces, jewelry, hairpins, needles, pocketknives and whips and sold them off with a fast-running patter.

"The fair used to last from Tuesday until Friday and came never earlier than September," said Bert. "There were no night fairs, because obviously you couldn't light up the landscape. The first day you'd just go to get your entries in. Wednesday and Thursday were the big days. On Friday you could buy fruit—peaches and pears—and get real bargains."

The first-day list of entries at the 1887 Fair was long:

Cattle in Class 1 (David R. Scott, superintendent) included Shorthorns, Devons, Holsteins (Friesians), Jerseys, Ayrshires, Herefords, Guernseys, and "natives." There was a herd prize and one for working cattle and steers. Times had changed since 1856 so there was a prize for "fat cattle." Class 2, for horses (Edwin Wakeman) included stallions, colts and carriage horses. Class 3, sheep (Archibald Tweedie) offered premiums for purebred South-downs, Leicesters, Cotswolds, Shropshires, Merinos, Hampshires, and Oxfordshires—and "fat sheep." In Class 4, swine (George Burrhus) could be found: Berkshire, Poland China, Chester, White, Essex, Yorkshire and Suffolk boars and sows.

The poultry entries included turkeys, white China and grey geese; white and black swans; Rouen, Aylesbury, Cayuga, Muscovy and Pekin ducks, as well as a great variety of chickens and Angora and Lop-eared rabbits, white fan-tailed pigeons, Jacobins and various other kinds of birds and pet animals.

The class for butter and cheese (Calvin McAllister) gave premiums for a firkin or tub of butter made "in the creamery" and "in the dairy" and for a "package of butter not less than 15 pounds."

There were 13 classes in all, with more than 1,300 entries. The other superintendents were Thomas H. Thomson, Farm Products; Charles Watrous, Fruits and Vegetables; W. W. Ogden, Farm Implements; W. B. Lyon, Mechanical Manufactures; E. P Berray, Discretionary; C. W. St. John, Boys' Department; and the Rev. Scott B. Rathbun, the Rev. O. D. Ramsay, Mrs. John Olmstead and Mrs. E. L. Guild, Ladies' Department.

Paintings, sketches and watercolors were shown as were quilts, clothing and embroidery. But also exhibited were things like fire screens, lamp mats, decorated pottery and carriage afghans.

Premiums ran anywhere between $1 and $4.

Membership tickets for admission were by gender. Men's tickets admitted two and were $1. Ladies' tickets admitted one and were 50 cents. These tickets entitled men to enter livestock, produce or other items; the women could enter items in the "Ladies' Department." A single admission was 25 cents, carriages, 25 cents, and "carriages belonging to holders of membership tickets admitted free."

The climax of the Fair was a Grand Procession of all the horses and cattle on Thursday morning and, in the afternoon, an address by Col. F. D. Curtis, editor of the *Farm Journal.* He discussed the practical side of farming, listed ideas for improving farm operations and suggested ways to increase both income and fertility. At one point he declared that "every farmer should keep sheep, however large his dairy..."

The Walton Reporter gave extensive coverage to the 1887 Fair in its issue of Saturday, October 8. It reported that the event "was all and more than it promised to be. When the week opened with rainy weather, it was generally thought that it would seriously interfere with the Fair, not only in the matter of attendance but also in the number of entries. But this drawback did not seem to discourage the people in the least.

"The second day was showery and disagreeable and yet the attendance was remarkably large to the surprise of everyone...It was found necessary to postpone the Wednesday attractions until the following day, and visitors therefore devoted their time almost entirely to viewing the exhibits. It is a difficult matter to do justice to the large and fine displays in each department, but it was generally conceded that the ladies took the first place in the line. The number of entries in this department was large and many of them valuable. The exhibits of paintings and fancy work were much admired, while in the more practical part of cookery and home manufactures, there was a very good display.

"The exhibition of poultry was large, and the lover of fancy poultry could find much to admire in this department. Richard Dunmore, of Franklin, led in the number of exhibits, having nineteen coops. H. R. Wheat, of Croton (Treadwell) came next with twelve coops of various breeds. . ." The Rev. Scott B. Rathbun, rector of Christ Episcopal Church, apparently was a priest who liked chickens; he is reported to have entered "a noticeable display."

"The fruit and vegetable department received much attention, and seldom has there been a finer display at a country fair. L. M. Fitch of Walton had the largest exhibit of potatoes and Henry Clark of fruit. The big Mammoth King squashes exhibited by William McDonald of Walton occasioned much comment. One of these weighed 153 pounds.

"The fear that there would be light exhibits of cattle proved groundless and this department contained some fine specimens of the various breeds. H. W. Dewey of Sidney Center had nine of his herd of Ayrshires on exhibition, while G. L. Fletcher of the same town had 13 fine entries of grades.

"The exhibit of sheep and swine was also good and many of these worthy of special mention. The exhibition of horses was not only large but included some fine stock."

In the Farm Trot for double teams John Wakeman and Leroy Smith were the winners. In the trot for single horses, winners were Walter DeLaney and the Allen brothers, and the five-mile go-as-you-please was won by Elmer Robinson. The Walton "Defiance" team captured a $50 first-prize in the afternoon race.

Walton merchants displayed items but did not sell them. T. Guild & Son had a large showcase filled with perfumes. Calvin DeBois showed watches, diamonds and clocks. T. L. Terry showed neckwear, Fitch Bros. & Seeley, carpets; J. B. Eells & Sons, "fine buggies and platform wagons;" Oliver and Ogden, coal and wood stoves; E. B. John, various kinds of stoves; O. J. Eells, willow rockers; Harvey Henderson, goods from his 5, 10 and 25-cent store; and the Novelty Works showed "a handsome line of samples of their manufactures." *The Reporter* suggested allowing goods to be sold another year—an idea that would provide nice receipts for fairs far into the future.

What was to become a popular adjunct to the Fair—the special O and W fair train—began its run with the 1887 Fair. It operated on the last day, leaving Livingston Manor at 8 AM. It arrived at Walton at 10, and left at 6 PM. One report has it that the train brought in 125 persons on that first trip and "if the weather had been better, it would have brought in many more." The Special ran for more than 30 years, usually on the Thursday, "sometimes consisting of 20 crowded cars." When more and more people began to own automobiles, the special train run became unprofitable. The last Fair special ran in September, 1923. The O and W had told the Fair board that the train would not run that year but apparently changed its mind. However, that was the last.

Finally, the newspaper's view of the 1887 Fair was that "it was a good-natured crowd that attended. In spite of the disagreeable weather, everyone seemed to have a good time and enjoy the show. No arrests were made, and the best of order prevailed."

Total receipts for the three days amounted to $1,347.12 which "places the Society in good shape."

In 1888 the Fair expanded to four days and attracted entries from exhibitors in many parts of the county. It was no longer just a local effort, and it offered three new features:

- A cattle auction
- A children's day
- A huge barbecue for which a whole ox was roasted all night over a charcoal fire.

The Fair was scheduled a week earlier - Tuesday through Friday, September 25 through 28 with John Q. Barlow as general superintendent and J. A. Churchill and P. M. Hanford assisting. The number of Society directors had increased to ten with one vice president instead of three: C. B. Bassett was again president, Joseph Harby, Jr., vice president; H. S. White, secretary and J. H. Townsend, treasurer. Directors serving three-year terms were the Rev. Scott B. Rathbun and Richard Garrison; two years, Alexander Tweedie and T. H. Thomson; one year, William C. Doig and William

C. Rogers. The Ladies' Department had grown. The Rev. S. G. Shaw and Miss Fanny Howland were superintendents, assisted by Mrs. W. C. Dole, Mrs. B. G. North, Mrs. William North, Miss J. C. White and Mrs. Robert Harby.

At the second fair premiums had increased—there was a herd prize for cattle at $10 and top premiums in several departments had risen slightly.

The Reporter put out an issue each day, thus becoming the first newspaper in Walton to be published daily.

The first day was again set aside for entries. Wednesday was designated as Children's Day when all county school children "who came in a body, in charge of their teacher" were admitted free. This innovation became a long-lasting tradition. In 2001, Wednesday is still Children's Day. The Grand Procession took place on Thursday followed by a guest speaker's address.

The auction, listed under the title of "Sales Department" was headed by the Rev. Scott Rathbun and was organized "for the purpose of bringing together buyers and sellers in large numbers." Friday was wholly devoted to this effort and it may have been successful, but by 1890 the auction had disappeared, an idea apparently hatched before its time but one which eventually grew and flourished, especially with the growth of 4-H.

In 1897, the fair board paid the Harby Brothers $5,000 for the tract it had been renting for ten years. During that time, many additions and improvements to the buildings had been made. The grandstand had a seating capacity of 1,500. By 1900 there was an Agricultural Hall with barns, sheds and stockyards. There were also buildings for the Ladies' Department, for vegetable, art, floral, domestic, boys' and poultry displays and space for equipment storage. At the 14th Fair in 1900, there were 6,100 exhibits and approximately $2,500 in premiums were paid.

Very likely a few automobiles were seen at the Fair in 1903 but essentially much remained unchanged. A. F. Reynolds was now president of the Agricultural Society with Robert Kilpatrick, vice president; P. S. McNee, secretary; W. A. Shepard, treasurer and W. R. Hall, C. A. Churchill, W. D. Potter and John McGibbon, directors. There were 17 vice presidents representing many surrounding communities, and the list of superintendents had grown.

Now, at the turn of the century, the simple ways were beginning to seem quaint; people were starting to crave something more than just bringing in farm products. Entertainers began to appear: the d'Atalies, "original French aerialists" doing "thrilling mid-air evolutions;" the Sigeberts "in their sensational comedy ladder act" as "aerial revolving meteors;" the De Forests who were "society jugglers" and "cannon ball manipulators" and who also presented a Dog and Monkey Circus.

It was about this time that the perennially popular Floral Parade began, possibly inspired by the Grand Procession which was so much a part of the first fairs. For many years businesses and local organizations competed to see who could create the most beautiful and flamboyant floats using all kinds of flowers in imaginative ways. These horse-drawn floats took many hours and considerable expense to produce, but when the fair board decided to give up the parade, popular sentiment kept it going, and for some years the parade continued. In the 1909 premium book there were 23 categories: There were society or club floats, and floats for businesses, firemen, schools, florists, farmers, truck gardens and printers. There were prizes for carriages drawn by teams or single horses, for bicycles, baby carriages, gentlemen and ladies on horse-back, for dog, goat and pony carts, and for four-in-hands and tandems. During the 1960s a kind of club parade was held for several years—but it was different.

Some fifty years later, looking back to the days of her childhood, Wilhelmina C. Greenmun wrote an article for the *Courier Magazine* of Deposit in which she reminisced about the Fair of 1905:

"All year around we young-uns looked forward to this particular occasion where we were even more gloriously happy than at Christmas or the Fourth of July. Shedding calicos, overalls and denim shirts, the youngsters got dressed up—in "flowered, flounced chintzes, or-gandies, or dimities with straw sailor hats and stiff-starched collars." Shoes! And neckties!

Their first view of the Fair "was the white tent-city filling us with excitement and anticipation; then the sound of the carousel filled us with joy, while it is impossible to describe the smell of sawdust, peanuts roasting, hot dogs cooking, clam chowder brewing, combined with the aroma of coffee brewed out of doors."

"As fast as possible we explored the marvels of the midway" and "rode ferris wheels, whip, merry-go-round, meandered in and out of side shows and watched games of chance. And we were always particularly attracted by the fortune-telling lady and the Minstrel Show."

At one o'clock everyone "assembled at the designated place, usually at the wagon one came in. There we sat on lap robes or horse blankets and passed around the finest food ever seen before or since.

"Out of the huge picnic baskets appeared sandwiches—chicken, ham, peanut butter, jelly and cheese—every kind of luscious cake imaginable, home-made rolls, cucumbers, ripe tomatoes and pickles, bananas and watermelon, all topped off by blackberry pies, jugs of cold tea or lemonade. . .

"After lunch the track horses began to warm up and the vaudeville acts took place on the platform directly across from the grandstand...while the Walton band played with such spirit that our old horse, Fan, felt like dancing —and did, too."

Two buildings were of special importance:

- The long, low building called "Ladies' Department" where "we saw on display all sorts of homemade dainties, pastries, candy, jellies, canned goods, needlework of every kind, quilts, hooked rugs, embroidered articles, clothing, art displays and flower arrangements," and
- "The other side of that building where the male population proudly showed their vegetables, fruits, chickens, rabbits and other poultry, as well as their hobbies—all in a spirit of friendly, neighborly competition."

The Munn Piano Co. gave away souvenirs—rulers, blotters, pasteboard fans and music books, Wilhelmina Greenmun wrote.

Next, she said, they headed for the stock tents "where even a more real spirit of competition was felt. Such fine specimens of cattle, no better in the East." She admired the horses, too, "all beautiful to see and wonderfully cared for. How proudly their owners brushed their coats and braided ribbons in manes and tails, for this was the era when horses were really necessary for work, for play, to get to church, for hauling produce to market—and for 'courtin'."

And then, the writer continues, "the wheels of progress turned and events tumbled after themselves so fast that time flew even faster: the World War, jazz, flappers, prohibition..."

World War I affected Walton mightily, as it did the entire nation, especially after this country became involved in 1917. But the Fair went on, reflecting the efforts for troops overseas and showing the productivity of the Victory Gardens. A new era was approaching and the Fair girded itself to cope with it.

Walton Historical Society Photo

Walton Historical Society Photo

A bird's eye view of Walton as it may have looked about the time the Delaware Valley Agricultural Society was making plans for its first Fair.

Walton Historical Society Photo

The covered bridge which had spanned the Delaware River since 1846 was something of a problem to those who wanted to organize a Fair in 1887 because of its age and general condition.

Photo courtesy of James Northrup

A contract to replace the old covered bridge was let in the summer of 1887, but, as often happens, the work was not finished in time for the Fair.

Walton Historical Society Photo

According to Murray's *History of the Century* there were 6,000 cows in Walton in 1897. Here are some of them, sometime around the turn of the century, crossing Townsend's Flat near where the school bus garage is today.

Walton Historical Society Photo

A closeup view of some of the early fair structures. The round tent may have housed the merry-go-round.

Photo courtesy of Helen Zandt

A barrel race required careful preparation and lots of assistance from the boys, some in derby hats and tail coats.

Photo courtesy of Helen Zandt

As soon as it became clear that the Fair would be successful, this grandstand was built opposite the judges' observation tower.

Walton Historical Society Photo

The fairground in 1897. This was the year that the Agricultural Society bought the land which it had been leasing from the Harby brothers. It paid $5,000. After laying out a half-mile track in 1887 the fair board went on to build, in the first ten years, an Agricultural Hall, a wooden grandstand, the Ladies' Department building and the stockyards. The judges' lookout tower at far left remained in that spot until trotting races were discontinued. The tower is now the judges' stand at the horse show ring. The first grandstand, extreme left, was replaced by a new one in 1907.

As early as 1892 the women of Christ Episcopal Church conducted a refreshment stand at the end of the Ladies' Building. Standing on the balcony are May Telford, Martha Lyon, Harriet Belknap and Flora St. John. Below, Claire Nisbett, Mrs. Charles Herring, Mrs. Nettie Burke, Mrs. John S. Eells, Miss Sarah Gardina, and Mrs. J. R. L. Nisbett, wife of the rector, and her baby, Anita.

Walton Historical Society Photo

Walton Historical Society Photo

One of the hottest attractions of the early fairs was the balloon ascension. Professor M. A. Allen, who described himself as "The World's Greatest Aeronaut..." was the first of several balloonists to thrill fairgoers for a number of years. The balloon, owned by W. D. Burns, was first seen at the Fair in 1896. It was inflated with hot air, it had a parachute suspended under the balloon, and a basket hung down below that. At nearly 1,000 feet over the crowd, acrobatic stunts would be performed on a trapese bar which dangled in the air.

Walton Historical Society Photo

Getting ready for the balloon take-off in 1900.

Walton Historical Society Photo

The Fair has attracted visitors from a wide area ever since its beginnings in 1887. In that year, a special O and W Fair train began its run to Walton and this "special" continued until automobiles put it out of business after 1923. The train started at 8 AM from Livingston Manor on the last day of the Fair and arrived at Walton at 10. It left Walton at 6 PM. At the station there were horse-drawn conveyances to take visitors the rest of the way. Boarding houses on South Street began early in the life of the Fair to offer overnight accommodation and some people put up at the local hotels.

Walton Historical Society Photo

Famous personalities visited the Walton Fair from time to time and in 1899 Theodore Roosevelt, who was then Governor of New York State, attracted a large crowd.

Walton Historical Society Photo

No, it is not raining. Umbrellas and parasols kept the ladies shaded from the bright sun in the early 1900s. In the foreground is an easel offering articles for sale, including parasols. Middy blouses were stylish when this was taken, about 1910, and of course everyone wore a hat.

Walton Historical Society Photo

The new grandstand, built in 1907, cost $2,179 and lasted for many years. Admission was 15 cents. Under the grandstand, behind the long row of windows, was a large exhibit area where organizations and local businesses could offer products for sale.

Walton Historical Society Photo

One of the Walton businesses to take space under the grandstand was the George O. Mead Company. Pictured in 1909 are Mrs. Rytmyer, George O. Mead, Cyrus B. Lincoln and Lillian Hess.

Walton Historical Society Photo

The late Mrs. Kennon Peck lent the original of this picture to be published by The Walton Reporter in 1961. Taken in 1910 or 1911, it shows the "new" 1907 grandstand in the background. The sign at right tells visitors to check bicycles and baggage at the round tent. The ticket taker in front of the sign is the late Adam Wolf. The wagon rig just to the left of the sign was a bus used to transport fairgoers from Delaware Street and the railroad stations.

Walton Historical Society Photo

A great deal of preparation—and pleasure—went into building the beautiful floats that took part in the Floral Parade, one of the favorite attractions of the early 1900s.

Walton Historical Society Photo

All dressed up, with flowers on their bicycles and wreaths on their heads, a group of girls—and a few boys—line up for the Floral Parade. Even the baby carriage is a bower of flowers.

Walton Historical Society Photo

In 1900, the first-place blue ribbon for decoration went to Lillian Neale, on right, and Evelina Peake McIntyre. Argie Neale, with his arms folded, is looking on.

Walton Historical Society Photo

Burns The Clothier, one of the early Walton merchants, had a salesman, Ward Thompson, who had a special gift for designing ingenious and entertaining creations for the Floral Parades. This float won first prize in 1902.

Walton Historical Society Photo

After winning the prize in 1902, Ward Thompson went on to design an even more inventive entry, a steamboat that moves along apparently under its own power. The secret is a man leading a horse, hidden under the prow. They provide the "motor" for the float, and belts, attached from the rear hubs to the side wheels of the steamer, makes them revolve.

—thanks to Dick Wakeman for the story

Walton Historical Society Photo

In 1908 Walton High School students, dressed as "Indians", rode past the grandstand on this leafy wagon.

Walton Historical Society Photo

Proof that some people in the early 1900s thought purebred poultry was "worth looking into" is furnished by this photo submitted by Mary Sherman some years ago. Her father, Clifford Fitch, was the driver.

Walton Historical Society Photo

An entire buggy covered with flowers was not an unusual sight in many Floral Parades. The fair board voted to end them in 1917 but they were so popular that they continued beyond that time and in later years updated versions appeared. Here, behind their sleek horse, are Mary I. Laidlaw Loudon and Margaret Kilpatrick in a photo dated about 1920.

Walton Historical Society Photo

Before the days of automobiles cattle were marched into town and on to the fairground. Here are half a dozen at the corner of Delaware and Bridge Streets, passing the home and store of Thaddeus Seymour St. John, the first president of the Village of Walton. They look quite different from the cattle of today.

Walton Historical Society Photo

Airplanes were another source of fascination for early fairgoers. To fly like a bird over the countryside and look down at the oval track and the tiny specks that were people was an adventure worth paying money for - and many did. This crowd was waiting for a ride in 1919.

Photo courtesy of James Kent.

By 1925 the construction of airplanes had changed only slightly.

Walton Historical Society Photo

Behind this smart-looking team are Mrs. C.B. Fitch and Mrs. A.J. Boice. In the background is one of the early dairy barns.

Walton Historical Society photo courtesy of Thelma Schriver

High-wire acts and animal performances were a great favorite of fairgoers in the early 1900s. These daring aerialists are "The Flying Dordens" who appeared in 1910.

Walton Historical Society Photo

Leopards riding bicycles? Yes, this and more when Valecitas' Trained Leopards appeared at in 1907.

Walton Historical Society Photo

A packed crowd was guaranteed when word got out that Lukens' Lions would be at the Fair in 1910.

Walton Becomes the County Fair

The Great Walton Fair in 1926.

It has been said that to have a successful fair, there must be an unending supply of dedicated, hard-working people who are in it not for themselves but for the whole enterprise, who want to see a growing, thriving agricultural county where the finest of farms and the finest of livestock are most prized. This was the thinking of the early fair builders, and it was a set of principles that guided the men who followed them into the often tumultuous 20th century when the Fair in Walton became the only thing of its kind in Delaware County.

The period following World War I was a time of unbelievable change. Railroads had developed to their utmost, to be threatened very shortly by Americans' love affair with the automobile; radio had arrived, television was a baby, and the next war would be fought in the sky. The great passion for agriculture that had characterized the mid-1800s, the "agricultural revolution," was waning. Farming was becoming less of a life and more of a business.

In Delaware County, its Fair was still firmly centered around farms and families, but no longer did one see farmer after farmer driving cattle through the village or entering the haywagon horses in a "farm trot." While family picnics at the Fair continued up until the 40s and 50s, the early years were times when families sat on the river bank with thick slices of buttered, homemade bread, tomatoes, home-canned pickles and a jar of lemonade. The carnival aspect of the Fair was growing, together with a realization that merchandise booths and various concessions, thrill shows and high-wire acts were increasingly what fairgoers wanted. A more sophisticated society was expanding the perimeters, a new kind of exhibitor, the breeder of purebred stock, was claiming attention, and with the development of 4-H, youngsters would be having a rapidly increasing role.

As far back as the mid-1900s, all kinds of sideshows had begun to proliferate, and food venders, and games of skill and chance. People began to expect more and more of a fair and it was becoming expensive. It was harder and harder to make ends meet. As the Fair entered the 20s, financial problems began to gather like thunderclouds.

Walton was not alone in facing a tough situation. In the early 20s most fairs were confronted with two vexing problems: Financial difficulties and the presence of "gypsies, con men and gamblers." An item in *The Delaware County Republican* in September, 1921 caustically noted that "the officers of the (Delhi) fair insist that there must be a big bunch of gamblers and half-naked human sharks (both male and female) in order to draw a crowd... There was a strong protest against the gambling games which occupied the midway—one experienced man estimating that there were 25 such 'attractions' on the grounds. But protests did not count this year, the church-men who conducted the fair claiming that all other fairs, except the State Fair, have these same 'attractions' including the shows which brazenly proclaim that their entertainments are suitable 'for men only', and they religiously adhered to the rule not to admit women no matter how strenuously the gentler sex insisted on seeing the sights." There were "scantily dressed women and men, games of chance, throwing craps and hoochy-koochying" (a kind of wiggle which probably inspired Elvis).

In 1922, *The Republican* reports, "Albany officials visited the Walton Fair...and found the grounds literally swarming with gypsies and depraved characters who were relieving the gullible of their money."

At the same time, there was another and dangerously pervasive factor affecting fairs: A creeping strangulation was beginning to overwhelm dairy farmers whose livelihood was being threatened by the creation of huge dealer combinations. Senator Royal S. Copeland would cut to the core of the matter at the 1924 Walton Fair. Capital and labor had learned the necessity for organization—farmers were not organized—and "only through better cooperation than now exists can the farmer secure rightful values for his produce," Senator Copeland would say. Two years later strong measures began to be taken to protect Eastern farmers from a threatened invasion by milk from the West and the advances of the dealer combinations. At one point in 1926, 125,000 dairy farmers held a mass meeting in Utica and voted unanimously for "unity," but as time was to prove, these efforts were in vain and only foreshadowed greater trouble in the decades ahead. But back to 1922.

Faced with what must have seemed an insurmountable fiscal crisis, the fair board, led by William H. Austin, then president, and J. Ward Palmer, secretary, called a meeting for November 25, 1922 at which they presented a plan to change the structure of the Society from a membership organization to a stock company. More than 15 members met in the Walton "Municipal Hall" and adopted four resolutions which

- Sanctioned the issuance of $15,000 worth of capital stock "to raise money for the purposes for which said Society was created"
- Specified that ownership of these shares would be the qualification of membership, and the holdings would "determine the extent of each member's voice in the affairs in said Society"
- Divided the stock into 300 shares having a par value of $50 each.

The certificate filed with the Delaware County Clerk is dated December 7, 1922.

Unfortunately this action did not stem the tide, since apparently not much stock was sold. As the fairground was being made ready for the 1923 Fair, scheduled for September 5-8, the news leaked out that the Society was $14,000 or $15,000 in debt. Stories were flying about that the Society might lose the fairground, that quite possibly there might be no 1923 Fair. Should this happen, it would be two years before state aid could be received again.

A proposition began to take form that the people of Walton might take over the fairground and use it for local activities such as picnics, reunions and community events. Some time in June, 1923, *The Walton Reporter* editorialized that "Changing times require changing methods and the fair needs a new birth.... It should not be allowed to die... A good deal is said about a 'purely agricultural fair' but the truth is that such a fair would not draw... Fair or no fair, the people of Walton will make the biggest kind of mistake not to get title to the grounds... It would make a playground such as few communities boast."

At a Chamber of Commerce meeting on June 25, a proposition was approved: That 20 business men and 20 farmers underwrite the 1923 Fair and manage it, and, because of the time factor, sell stock later.

By early August, however, the fate of the Fair was still up in the air. Of the 40 necessary guarantors, only 28 had come through. However, within a week the total had reached 30 and *The Reporter* announced with a flourish: "Will Hold Walton Fair—horse and auto races and platform attractions daily." The merry-go-round, ferris wheel, seaplane swings and the whip all would be there—and, of course, the animals.

So, the 1923 Great Walton Fair took place, and its board then pulled itself together and tried to sell stock. They were not altogether successful.

Meanwhile, the County Fair at Delhi was taking place from August 28-30, an earlier than usual date, ostensibly to avoid conflict with the State Fair, but probably also to get an edge on Walton as well since the two communities had always competed strenuously for fair patrons. While Walton wrestled with its financial problems, the organizers in Delhi were faced with serious decisions of their own. It is more than likely that they, too, were facing a financial downside. Soon the Delhi fair would disappear, and Walton would become the County Fair—but not yet.

Early in January, 1924, the Walton fairground was sold at public auction.

Some 40 persons attended the January 5 sale in the office of Samuel H. Fancher. He was the referee in foreclosure proceedings brought by the First National Bank of Walton, as trustee, in order to satisfy a second mortgage which had been granted to cover some notes.

The bidders were A. J. Courtney, Village President; Dr. W. B. Morrow, W. H. Austin, president of the Fair board; and Lawrence D. McClenon, a director. After slow and cautious bidding which started at $5,000, the 30-acre fairground was knocked down to Mr. Courtney for $12,000. Purchase by the village president "assures the continuance of the Fair and the use of the grounds for amusement purposes," announced *The Walton Reporter.* "A number of Walton business men stand ready to enter a stock company with Mr. Courtney."

Within a month Jerome J. Farrell and E. B. Guild had joined Mr. Courtney and Dr. Morrow to form a committee. It immediately set about selling the recently authorized stock. "Those familiar with the situation are optimistic," said *The Reporter,* that a fair properly conducted could "show a substantial profit and permit a dividend to the shareholders." Nothing more was said about townspeople's picnics, reunions and community events at the fairground.

At a special meeting on June 16, J.J. Farrell was elected fair board president; Dr. W.B. Morrow, vice president; Earl S. St.John, secretary, and Henry S. Ogden, treasurer. Elected as directors were H.D. Henderson, Arlan J. Wood, Peter Chambers, A.J. Courtney, Charles Wheat and Otto J. Hoos. Charles Wheat was named grounds superintendent.

Full speed ahead to the Walton Fair of 1924, September 2-5. Its spirit had been renewed, its premium list revised—upwards—its racetrack gone over, and a special effort made to get a "forceful and pleasing" speaker—Senator Copeland.

For once, sunny weather favored the Fair and it attracted "the largest crowd in recent years." They had to stop selling seats in the grandstand; it would only hold 1,200. Autos were parked all over and state troopers directed traffic. There were two harness races daily for three days (purse, $300), daily "aeroplane" flights, a parachute jump from the plane, and State Police Troop C of Sidney giving its popular horsemanship program.

Booths were under the grandstand at this time; the Grange's exhibit of vegetables and farm produce was particularly admired. Other 1924 booths were those of the anti-tuberculosis association, the high school, J. E. Wood's stoves, Camp Milling Co., the First National Bank and the Farm Bureau. Junior Project workers had their own department; demonstrations were held by various units.

There were "exciting" platform attractions including fancy dancing by the Caparula Sisters and an unusually large number of horses in the races. Senator Copeland, speaking in front of the grandstand, stressed the necessity for better organization among farmers. The remarks came during a year when Sheffield and other milk dealers had instituted a price-cutting policy which flattened the market.

Also, the fair association had "barred all questionable shows and their attendant games of chance," after which *The Walton Reporter* noted that "there are fewer side shows than usual but the refreshment stands are numerous." The change brought applause both from the ministerial society and citizens at large.

A week later, at the Delaware County Fair in Delhi, National Defense Day was observed with a parade, speeches and exercises. Tuesday, the opening day in Delhi, was spent arranging exhibits, but Wednesday was so cold and rainy that the Fair did not get into full swing until Thursday. According to all available accounts, it had just one more year to go.

During the Spring of 1925, the horseracing track in Walton was completely rebuilt by Coates Brothers of Goshen and was pronounced to be "one of the best half-mile tracks in the state." A new guard rail was built around the inside of the track. To increase the seating capacity of the grandstand, bleachers were built so that more people could enjoy "Paulus the Peerless" on a trapeze 75 feet from the ground, a group of acrobatic clowns "par-excellence," the Jim Jam Jems; and the Rosa Rentz Duo, a ladder act. "Healthy," the Dairymen's League clown, was to be on tap and horse races were scheduled two a day, Wednesday through Friday. The new Walton Band played every day. The premiums to be offered totaled approximately $6,000.

Although rain washed out one day completely, causing events to hold over until Saturday, the 1925 Fair, September 1-4, ran in the black.

"The Fair can be made to pay!" *The Reporter* exulted.

The fair association had been wise that year; it had insured itself for $3,600 against rain. Gate and grandstand receipts topped those of the previous year and concessions brought in more money than ever before. Expenses had included $4,335 for premiums, $2,036 for attractions and music and $1,598 for horserace purses. At the final reckoning it appeared that not counting the building and grounds improvements, the Fair showed a $1,490 profit. Trouble was, however, $1,325 worth of the stock was still not paid for.

Meanwhile, once again in 1925, Delhi, which usually scheduled its Fair perhaps a week after Walton, decided to hold it three weeks earlier, August 11-14. Whatever the hoped-for gain, the move failed.

Early in 1926, *The Delaware Republican* reported that the Delaware County Agricultural Society might disband, officers of that group having asked their attorney to look into the process of dissolution. Apparently officers for 1926 were elected, but available records never again mention the Society or the County Fair in Delhi.

Whether financial problems were crushing Delhi, or whether the general crisis facing dairymen cut into its attendance and number of exhibitors; whether its Midway had become a commercial elephant dwarfing the agricultural aspect of the Fair, or whether the earlier complaints and protests against the gambling games had severely damaged that Fair—or whether it was all of these things together—is impossible to tell from the scanty records that exist.

George Stewart, the well-known proprietor of Stewart's Department Store on Main Street, Delhi, is a man whose memory goes back to the early 1920s, when he was a small child. Why the fair dissolved, however, he could only guess.

The family lived not far from the fairground, and, he says, he was never allowed there unaccompanied because of the large number of gypsies, who, in those days, were regarded with awe, if not fear. He notes that while Walton rid itself of undesirable features, Delhi possibly did not, and there is no doubt that money was tight. Furthermore, there was no room for expansion to speak of. On the north was a swamp and a canal, and the Delaware River and the highway hemmed in the east and west boundaries.

George Stewart has been a member and officer of the Donald W. Gleason American Legion Post #190 for many years and therefore is well acquainted with the situation after the Legion bought the fairground in 1929. His feeling about the demise of the Fair is that "I think they were having problems and were glad to turn it over to someone else."

In an article written some years ago, Mark Boshnack quotes a *Delaware County Dairyman* editorial of August 23,1929:

In recent years, agricultural fairs have had a hard time.... Many of them, instead of keeping to their original purpose, became too exclusively an entertainment show with Midways and Vaudeville.... The Walton Fair is our one remaining agricultural show, and it is becoming a real country fair. The management has sought to keep it balanced with enough entertainment without sacrificing its features as a farm show.

The year 1926 began with more unity talk among the dairymen, and among the horse people there were plans for a tri-county circuit which would organize the racing schedules into a consecutive series of dates. Afton, Norwich, Oneonta, Cooperstown, Morris and Walton were on the circuit.

The dairymen met in Utica in March and formed a "committee of 11," charged with producing a "plan of coordination," but critics of this move complained that "the fact that the committee is composed of officers of the different associations preconceives failure... It was a mistake that the Utica meeting did not order that the committee be composed of farmers, not of officers of any association." History was to prove this an accurate assessment.

The Delaware Valley Fair (as it was called by *The Reporter*, (just once) was moved from September back into August (Aug. 31 - Sept. 3). As the dates for the Fair approached, *The Reporter* editorialized:

"People come from long distances to attend the fair. One party drove in on the most rainy day last year in an old Ford in which they had traveled 40 miles, covering the last eight or nine without fan belt or air in one of the front tires... No information was obtained as to whether they ever reached home again. If some people will go to all that trouble to reach this great occasion, there can't be any doubt that everyone who can ought to be in town each day of the fair."

In its advance articles, *The Reporter* played down the agricultural aspects and emphasized the "bigger and better" free attractions:

- Tice's Dogs, a dog circus employing 12 cute pups.
- The "Fearless Greggs," the "most expensive act that booking houses carry"—two autos, one containing a woman, the other a man, rush down a steep incline at top speed. The first car leaps into the air and turns three complete somersaults while the second passes under it and lands first.

The harness races were predicted to make "all previous years' track events appear as evening buggy rides." The Walton band would play and Troop C would give its perennially popular exhibition. Yes, the animals would be there, too. For the first time all cattle had to be tuberculin tested.

When the 1926 Fair was over, reports indicated that 1,000 Junior Club Workers (4-H) had been involved, the largest number to date. The cash statement, taking everything into consideration, was about the same as the year before.

In 1928 the Fair advertised, among other attractions, two in particular: Four trained military elephants ranging in age from 40 to 110 years of age, including one that had "carried thousands of children on her back" and was "careful as a mother can be," and a prominent guest speaker, former Congressman and past president of the New York State Farm Bureau, Peter Ten Eyck of Albany. His topic was to be "Cooperation," the most serious subject of the day for dairymen who were trying to bring together the milk producers in order to improve distribution and get a more adequate return.

The elephants came and were highly praised; the Congressman was not mentioned again.

Apparently it was a nice Fair: "the midway this year is the largest it has ever been" and some 1,600 cars were clocked into the fairground on Thursday. But a rainstorm put an end to Thursday's festivities and some horse races had to be run the following day. There were more exhibits in all departments except art, poultry and flowers. Guernseys, Ayrshires, Jerseys and Holsteins were entered in the cattle division; the Ayrshires carried off the prize for best herd, followed by the Holsteins. Sheep and swine entries were numerous and exhibits in the ladies' department "were up to their usual high standard." A dahlia show, planned for fair week, was postponed because wet weather had held back the blooming season.

After the annual stockholders' meeting in December *The Reporter* carried bad news: "Walton Fair Lost Money This Year."

In 1927 the Fair had shown a slight profit. Although in 1928 gate receipts were greater, the "attractions" were more expensive and premiums had cost $700 more. Another $600 had been spent to upgrade the buildings. Henry S. Ogden, treasurer, reported a deficit of "approximately $1,200."

In the August 17, 1929 issue of *The Reporter*, the headline read: *Fair Is County's Old Home Week.* "Many years ago each of the villages in the county had their fair" (a slight exaggeration), "but they have been gradually discontinued until now the fair in Walton is the only one left and the residents of the county now regard it as their fair and come from far and near....The week of the *Walton Fair* (italics ours) has rightly been called Delaware County's old home week as it is during that time that the people of the county congregate here and renew old acquaintances..."

What was happening was that the Fair was sliding gently into its new role as County Fair, not making a bold statement but by its continuing presence establishing a fact.

At the very end of a *Reporter* article on August 16, 1930 a newsman quietly wrote: *This is Delaware County's only fair. Make it your fair. Plan to come and bring your family and friends.*

According to the state Department of Agriculture and Markets, the title of "county fair" is not an official designation, and most county fairs are run by groups that have the words "county agricultural society" in their official names—which the Delaware Valley Society did not.

Because they were aware of this, the 1920s fair board may have been reluctant to announce itself immediately when in 1926 the fair in Walton became the only one in Delaware County.

But it had won an important competition. It had proved itself the best by operating an agricultural exhibition of high quality and by putting on an accompanying carnival that was, as many have insisted at that time and since, "a clean show." It had integrity, forthright management and a devotion to the original goals. It was, indeed, The County Fair.

To be sure, the Fair had, and would continue to have, a roller-coaster ride of good years interspersed with hard times, but somehow there were always leaders to be found who would pull things together. The most celebrated of these, George Rich, who led the Society for 33 years, is said never to have referred to "The Walton Fair." It was always *The Delaware County Fair* because it was for the whole county, all of its people, and a celebration of the best in central New York State.

Walton Historical Society Photo

Photo courtesy of Helen Zandt

With the ferris wheel and exhibit tents in the background, a farm buggy arrives at the Fair with its spruced-up team of horses.

Photo courtesy of Hanford Mills Museum, East Meredith, New York

Back home, haying with a troika, (a team three abreast) was not unknown.

Walton Historical Society Photo

Early in the 1900s, this was the E.E. Rissley farm on Third Brook, with its splendid four-storied barn and large chicken house. Later it became the property of the William Austin family, famous for its cattle auction business on Bridge Street. Bill Austin was president of the fair board for a number of years during the 1920s.

Walton Historical Society Photo

Inside the Austin auction barn a crowd gathers to look over the latest shipment of horses from the West. Many a horse, cow, bull or steer went to local farms from Bill Austin's auctions, and from there to the fairground to enter the livestock classes and compete for the premiums.

Walton Historical Society Photo

The Townsend Farm, at Howell and Delaware Streets, was another spot during the 20s where large numbers of livestock could be found.

Walton Historical Society Photo

Farm machinery has always been a point of interest with the organizers of the Fair and its patrons alike. J.E. Wood of Walton not only sold stoves, he was also involved with the latest in mechanical developments for the farm. Here is a group displaying a new hay blower, probably some time in the 30s.

Ticket courtesy of Helen Zandt

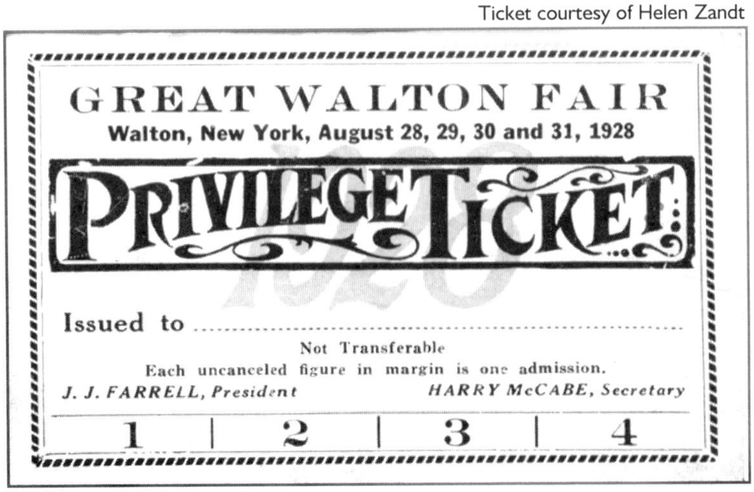

A 1928 ticket to the Great Walton Fair.

The Years Fly By

The thirties brought hard times to Delaware County and the forties began with War. Through it all the Fair carried on, offering each year's crowd a bigger and better program of "amusements, exhibits and entertainment." It was a bumpy ride and the fair board must have been glad to see the 50s arrive. But the up-and-down pattern of make-money-one-year-lose-it-the-next was beginning to take its toll.

During the first few years of the 30s the Fair, which was still calling itself The Great Walton Fair, pretty much followed the pattern of the preceding decade. Electricity having come to rural New York (although not yet to most farms), floodlights were now possible, and in 1930, for the second year, there were night fairs on Tuesday, Wednesday and Thursday, with the ever-popular fireworks.

During the day there was horse racing to watch, "hundreds" of farm exhibits to examine, and Captain Daniel E. Fox of Troop C, Sidney, with his "internationally famed riders," had "many new tricks plus all the old ones."

Miss Marlette "of Ringling Brothers Fame" dropped from "a miniature tin box hoisted high in the air." Other trapeze artists and circus riders were there, together with a contortionist and a trampoline act. In its follow-up story, *The Walton Reporter* asserted that "the Fair has been improving steadily in later years and the 1930 one can be said to be the best ever."

After the Fair in 1931, however, a reporter labeled it "one of the most successful seasons as far as exhibitions are concerned and one of the poorest in regard to attendance that has ever been had."

Infantile paralysis had been reported in Downsville, followed by a false rumor that no children under sixteen would be allowed at the Fair. And then, on the most popular day, Thursday, it poured; at night, after the fireworks display, all the lights went out. Consequently, attendance was down "about 65 percent," and the fair board was left looking at "a large deficit."

There was a census report in 1931, too, which indicated a sharp drop in the number of farms in New York State, but, surprisingly, an increase in the number of cattle. The number of farms had dropped from 193,195 in 1920 to 159,806 in 1930, with their corresponding value decreasing from $1,425,061,740 in 1920 to $1,315,904,741 in 1930. The number of horses and mules had decreased significantly, but cattle numbered 2,144,244 in 1920 and 2,220,139 ten years later, an increase of 75,895 animals.[1] The loss of family farms, of course, has continued. According to a report in 1999, which surveyed a 10-year period, the number of farms in Delaware County declined by 33 percent between 1982 and 1992.[2]

In 1934 the Fair formally took a new name.

As the premium books arrived, people noticed something different. There, on the cover, were the words:

The Delaware County Agricultural Fair and Horse Show.

This title remained in use until the change in management in 1957.

The Mighty Reithoffer Shows arrived in 1934 also. It was a family-owned venture started before the turn of the century, and it was to become as much a part of the Delaware County Fair as the merry-go-round was a part of Reithoffer.

Over the years the children and grandchildren of Julius Reithoffer, the founder, were to prove so flexible and creative at managing the business, having as their goal "a good, clean show," that in 2001 Reithoffer's Rides, is celebrating 67 years on the Midway.

The night fairs attracted increasingly large crowds, and there was horse racing "for some of the largest purses offered in this section of the state." In 1934 and 1935 these were running races, however, and they marked the beginning of the end for horse races in Walton, although harness races did come back, and they continued off and on until the middle 60s. About this time an additional horse barn was built.

President of the fair board was Peter Chambers, who had been elected in 1931 and would serve for 25 years, one of the longest terms in the history of the organization.

In 1935, the Midway was moved to its present position in back of the grandstand. It had always been located within the track along with all the tents and buildings, but it was growing and competing for space with a new baseball diamond, also gaining in popularity and attendance. In its new location, there could be almost no limit to the Midway's growth. The change also opened up more space for parking inside the track.

The Fair of 1935 featured a Delaware County historical pageant dealing with the anti-rent war which had so bitterly divided the populace from 1838 to 1849. Members of the cast came from six communities. Walton was not on the side of those protesting the feudal rent system, according to historical accounts.

Captain Fox's troopers from Sidney were there again, and heavyweight world champion boxer James J. Braddock's wife was Fox's guest one day.

The entertainment that was so popular then—aerial acts, a clowning cyclist, a trained seal, juggling, perching and balancing performers, and a Funny Ford which danced the Charleston and chased its owner—gave their performances on the "roof garden" in front of the grandstand.

In the agricultural department there were 13 sections. A notice advised entrants that since this was "a strictly amateur exhibition" classes in most cases (cattle and horses excepted) would be closed to all but Delaware County residents. In the past, "professional exhibitors" had competed and "took away large prizes," *The Reporter* explained. There was a wood-chopping contest, the Walton band played and there were two nights of fireworks.

In addition to all of this, a bit of unplanned entertainment centered around the escape of Jerry the Monkey. On opening day, as his owners were getting ready to open their concession, Jerry, who was a large monkey, broke his chain and headed for the tree tops along the river bank. Everybody thought Jerry was cute but he turned out to have a miserable disposition, threatening to claw anyone who came too close. Baring his teeth, he dared all comers to capture him, and in a "catch-me-if-you-can" performance, he appeared next on Stockton Avenue, and two days later in a small tree near the Stockton Avenue armory.

The men in charge of Jerry apparently were not awfully anxious to go up the tree after him, and they went back to the fairground to get a trap. But while they were gone, Jerry scampered down the tree and into Mrs. Jenks' garden, where he disappeared for a while only to reappear in another tree. The trap was set at the bottom, but the monkey ate the banana that was in it and leered at the crowd. Only a ripe apple at the end of a pole and the ingenuity of his keepers made his capture possible. They hooked a wire through the chain which he still had around his neck and dragged him off.

Without a great deal of hard work and devotion, this successful Fair of 1935 might not have happened. On Monday, July 8th, a day that would go down in local history, the fairground (as well as many sections of the town and county) were devastated by the worst flood in memory. Water went rampaging down the main street of the village and at the height of the disaster, a fire destroyed the Melnick Store on Delaware Street. Long hours of clean-up work awaited store owners and town crews. In some places streets were buried under four feet of gravel, and washouts had occurred all over.

At the fairground it was not only a matter of repairing the damage, but of doing it in time for the opening day on August 27. Deep water-filled trenches were to be seen near the grandstand, and the track had to be filled in and freshly graded. Legend has it that the holes were so deep that they swallowed automobiles. History would repeat itself in 1996, but a few who remember the Flood of '35 declare that one to be the winner in terms of devastation and ruin. Besides, heavy equipment was not that "heavy" in 1935 and a great deal of hand digging and hauling was necessary. It was a dedicated and joyful crew that by the first weeks in August announced: "We can have a Fair!"

During most of the 30s, any tranquillity there might have been on the farm scene was shattered by continuing dissension in the milk shed. 1935 was characterized by earnest, but largely unsuccessful, efforts to "stabilize the dairy industry." A milk shed conference had been formed consisting of:

- Representatives of general farm organizations,
- State colleges of agriculture,
- Milk control boards,
- Producers' organizations, and
- Distributors.

These had been meeting diligently in an effort to work out a milk marketing agreement fair to all parties. But all parties could not agree, and the stage was set for the milk strike and violence of 1939.

As far as the 1939 Fair was concerned, the milk strike was badly timed. The Fair was scheduled for August 22-25. On the ninth, some 250 farmers attended a county-wide mass meeting of the Dairy Farmers' Union at the court-house in Delhi and voted 235 to 6 in favor of a milk strike. What they wanted was $2.35 per hundred-weight for 3% milk in August, September and October. The price had been running around $1.50.

Five days before the Fair was to begin, striking farmers descended on the milk plants "in one of the most strife-ridden chapters in the history of dairying in Walton."[3] Stones, hammers and clubs were wielded as strikers tried to prevent the milk haulers from delivering cans to the creameries. At the Dairymen's League Creamery in Walton, M. K. Hunt of Mundale watched as pickets destroyed half his load. Another victim was Hiram "Hi" Barringer of Franklin whose truckload of milk cans was dumped in the street, cans flying in the air and milk running down the road like a white river. Tempers ran record high and who thought about a Fair?

But almost as quickly as it had flared up, the strike was settled, and on the Tuesday following the violence, crowds trekked into the fairground to enjoy another year's festivities. Afterward, however, *The Walton Reporter* noted that "Curtailed pocketbooks caused by drought, the distraction of a milk strike and, finally, a heavy downpour, just as the 'big day' (Thursday) was getting nicely under way, conspired against the 54th edition of the Delaware County Fair."

Later reports set the gate receipts at $1,200 below the previous year, but apparently fair directors had anticipated a poorer season because of low milk prices and had reduced expenses. "This allowed the fair association to end its fourth consecutive year with a small cash balance after paying all bills."

With the 40s, the shadow of war in Europe began to fall over American life, but as before the repercussions took longer to reach more remote areas like the western Catskills. The Delaware County Fair of 1940 offered "milking contests, horse pulling and wood chopping contests, running and trotting races" in addition to the usual horse, cattle, sheep, and swine sections.

The "free amusements" at this time were being supplied by the George L. Hamid organization whose business it was to provide all kinds of acts to a long list of fairs. Trained steers, prancing dogs, high wire artists, contortionists, and trained lions and tigers had been entertaining fairgoers since the early 1900s. With the arrival in 1940 of Lucky Teeter and His Hell Drivers, however, it could be said that the climate had changed. People's interests were running in a new direction, thrill shows, which, after the war, would claim top billing.

During the War, the Fair was cancelled for only one year: 1942, because of travel restrictions. The 4-H set up Community Victory Harvest Shows and a Junior State Fair was held in Cortland County. At that time there were 4,267 Victory Gardens.[4] In 1943, the early word was that the Delaware County Fair would be cancelled again, but when the Office of Price Administration clarified its "unnecessary driving" edict to allow farmers and fair exhibitors to use their cars, the Fair hurriedly scheduled itself for September 21-24, the latest date in many years.

Special emphasis at the 1943 Fair was on farm exhibits, 4-H projects, the foods-for-victory program and displays for the sale of war bonds. *The Reporter* remarked: "Exhibits should be better than if the exhibition were held in August, as farm crops will be nearer maturity."

In 1944 the polio epidemic may have curtailed some of the youth activities. John Merchant served as 4-H agent during that year, a brief successor to the county's first agent, John Lennox, who had become the state 4-H leader at Cornell. Early in 1945 Harold Carley arrived, the man who was to do so much later to save the Fair.

The 1945 Fair featured a Pet Parade open to all youngsters under fifteen. Dogs, donkeys, ponies, chickens, goats, rabbits, ducks, parrots, pigs, mice—all kinds of pets were eligible to compete—except cats! At that fair a special guest was Gov. Thomas E. Dewey.

As memories of the war years gradually faded and life for everyone became more normal, more and more out-of-town visitors made the August fair week their vacation time. Many years earlier a boarding house had been established somewhere along the Fair route, and the Walton House and the Austin Hotel did a good business. Now with increasing numbers of visitors, the houses along South Street took advantage of their proximity to the fairground, and a number of them were offering people a place to stay.

These were the years, also, when the involvement of farm and home organizations reached a peak. Twelve Granges created exhibits for the 1949 Fair: Pepacton, Mundale, Walton, Hamden, Hancock, Andes, Delhi, Franklin, North Franklin, Harpersfield, Stamford and Mason--ville, together with five Home Bureau units from Kerr's Creek, Downsville, Merrickville, Franklin Depot, New Kingston and Sidney.

A Better Milking Contest, which had begun several years earlier, was sponsored by the county Farm Bureau with Ayrshire breeders furnishing the cows and various farm equipment dealers lending the equipment. Members of the Future Farmers of America and 4-H were eligible to compete. The Farm Bureau had an exhibit on mastitis control. The Walton Artificial Breeders Association also had a display of this reproduction method.

As the 50s began, the Delaware County Fair had become known far and wide. Its program and exhibits were doing fine, the part played by the 4-H was becoming a source of pride and attendance was increasing year by year. The financial statements, however, were beginning to show a somber picture. In 1952 the Fair reported a net profit of $185.98. In 1953 that figure was $313.89. In 1954 it had a loss of $2,406.87 and in 1955 it lost $133.67. Something, sooner or later, was bound to happen.

This was the grandstand being used in the early 1930s.

Walton Historical Society Photo

Sometime during the 30s, Franklin Delano Roosevelt visited the Fair.

Walton Historical Society Photo

A swirling torrent threatens to destroy the track in front of the grandstand during the Flood of 1935. Water lapped at the edge of the bleachers, damaged the fence, and left craters so large that they are still a subject for tall tales around the village. The flood, on Monday, July 8, was caused by heavy thunderstorms and pelting rain which in a short time taxed all of the small steams beyond their capacity. It was only six weeks before opening day of the Fair.

GREAT WALTON FAIR

September 1, 2, 3, 4

Night Fairs, Wednesday and Thursday, Sept. 2 and 3

Both Educational and Entertaining

Better exhibits, better racing, better amusements than ever before

FREE ATTRACTIONS

FOUR FANTINOS

Acrobats of the European school now touring this continent. They not only feature aerial acrobatic stunts but also the great strength of Miss Mimi Fantino.

EL RAY SISTERS

Known as the dancing roller skaters. Anyone who has ever used roller skates can appreciate this act and will marvel at their ability to keep their feet while going through their intricate steps and stunts.

Horse Racing Wednesday, Thursday and Friday

ARTHUR & NORMAN

Fair time is a period of festivity and mirth and this act will cause even the most hardened cynic to split his sides with laughter. Two contortionists dressed as frogs, and their antics never fail to draw applause.

BELLECLAIRE BROTHERS.

Feats of gymnastic strength that excite wonder. A thrilling loop the loop into a back handed catch by the understander has provided thrills for the audience wherever this act has been staged, in this country and in Europe.

Capt. Fox and his Troopers will give an exhibition of horsemanship each day

AGRICULTURAL EXHIBITS

One of the main features of the fair has always been the exhibits. The 4-H Club entries have given a splendid increase of interest. County agent Lenox states that the boys' and girls' Department will be larger and better this year than ever before. In addition the best herds of dairy cattle in this section will be on exhibition. Over 300 head of cattle were judged last year.

Remember the dates:

Tuesday, Wednesday, Thursday and Friday

September 1, 2, 3, 4

A poster advertising the 1931 Fair bears the title that was used for many years, even after the Fair in 1934 officially began to call itself The Delaware County Agricultural Fair and Horse Show.

Photo courtesy of Jim Northrup

Originally, the Midway was located inside of the track, along with all the other exhibits and concessions. In 1935 it was moved to a spot behind the grandstand where it had room to grow. By 1948, its tents, rides, merry-go-round and merrymaking crowds looked like this.

Photo courtesy of Jim Northrup

For many years, the horse show ring was opposite the grandstand, as seen in this picture taken in 1948. At that time, virtually all of the exhibits at the Fair were under tents.

Photo courtesy of Jim Northrup

From the earliest times until the present, the number and quality of dairy cows and beef cattle have testified to the predominance of dairying in Delaware County.

Walton Historical Society Photo

The absence of farm machinery was ostensibly one of the reasons for the poor showing the Fair made in the middle 50s, but here, in 1948, the display was large and varied.

Photo courtesy of Thelma Shriver

"Death Defying Hell Drivers" have been entertaining fairgoers for as long as there have been cars. Lucky Teeter and Joie Chitwood are two names that leap from the pages of fair history.

From the Cornell Cooperative Extension Archives

E.B. Guild, president of the National Bank in Walton, purchases the first bond in the 1943 War Bond Drive, from Adrian Mostert of Delhi, a 4-H member of the Spring Peakes 4-H Club.

From the Cornell Cooperative Extension Archives

Gov. Thomas E. Dewey visited the Delaware County Fair in 1945. He is shown inspecting the 4-H dairy project of Leland Neff.

Walton Historical Society Photo, Courtesy of Thelma Schriver

Trotting races were part of the scene from the beginning, but eventually they disappeared.

1957: A Time of Crisis

In the early days, the line of horses and wagons tied to the racetrack railing testified to the eagerness with which families took the day off from field work and chores to head for the fairground with their picnic baskets and enthusiasm. Later accounts of the traffic congestion on Delaware Street during Fair Week reflect an increasing popularity as the Fair grew, and, to all appearances, flourished.

But all was not well. Financial problems had dogged the fair board from time to time, with a critical point occurring in 1924 when the fairground was actually sold at auction and then redeemed. In 1957, the Fair faced, for the second time, a year of crisis. The Fair all but ended once again, and the fairground nearly slipped into the hands of the City of New York for a sewage disposal plant. This was to be the story everyone would remember for its heroics, its clash of personalities, the antagonism and bitterness, and the role of certain farmers and 4-H leaders.

To deal with its problem of recurring indebtedness in the 1920s, the Delaware Valley Agricultural Society had reorganized, replacing a membership format with a stock company. It issued $15,000 in capital stock. But this move proved not to be sufficient, and an effort to save the Fair was mounted by a group of business men led by A. J. Courtney, Village President; W. H. Austin, a cattle dealer and president of the fair board; Lawrence D. McClenon, a director; and Dr. W. B. Morrow. At the auction, the fairground was hammered down to the village president for $12,000.

After the businessmen took over in 1924, the Fair was able to keep going more or less well, depending upon the state of the economy, the local dairy situation, and the weather. Rain, all through the years, has been a plague and a discouraging money-losing factor. As for the economy, the 30s could hardly have been worse, followed by World War II which, at least in 1942, brought everything at the Fair to a standstill.

During these years, a new set of names began to make their appearance, including that of Peter Chambers, a breeder of pure-bred Holsteins; Paul G. Williams, who ran a fleet of school buses; Henry S. Ogden, insurance broker; and C. O. Field, high school principal. These men were still serving in the middle 1950s, and they had been through one or two grilling times. By then, the Fair was again perilously in debt. When it lost money once again in 1955, people who had bought stock at $50 a share were beginning to be very worried about their investment.

These investors ranged from the largest stockholders, the First National Bank of Walton with 10 shares and the estate of former fair board president and cattle dealer William H. Austin with 6, to fair board members and other people in the community, including 4-H members and clubs, some of whom owned as little as one-fifth of a share. And, as one of the board members admitted later at one of the more torrid community sessions, the Fair officers seemed to have become "tired of running a constantly losing enterprise." Their solution was to try to sell the fairground—twice.

It should have come as no surprise when early in 1956 the Walton Central School Board of Education, which was working on plans for a new high school, announced that it had taken an option to buy the fairground for $25,000. This sum would have paid off the Fair's debts and enabled the stockholders to retrieve their investment.

During 1956, the school board and its architect, Perkins and Will of White Plains, investigated three sites: The fairground with the Merritt Loudon farm added to it; a combination of lands owned by Joseph Harby, Dr. Robert V. Schatkin and Mrs. Julia Gregory on the hill above Stockton Avenue; and the Neal greenhouse property on East Street. By the time March had arrived, the board architects were advocating the Fairground/ Loudon tract as the best site, and in June, at a meeting attended by 250 persons at the Townsend School, this proposition was hashed out.

A reporter later wrote: "Brisk questioning after the architects presented their side of the case revealed that there is a large element which does not agree with the architect upon his choice of the fairground..." In particular, observers commented upon the danger of flooding. One man, who had worked at the fairground for 50 years, said he had seen many floods, including the one in 1935, when torrents poured in around the grandstand and the area had to be closed. Supervising Principal Thomas H. O'Neill told *The Walton Reporter* that people were hailing board members on the street and asking them questions.

At this time the fair board was composed of Peter Chambers, president; Wallace Rich, a poultry farmer from Hobart, vice president; E. B. Guild, a druggist, of Walton, treasurer; Mr. Williams, secretary; and Mr. Ogden, Mr. Field, Louis Camp, feed mill owner from Walton; Ralph Higley, president of the First National Bank, Walton; Carlton Chambers, Walton, and William Gifford, a farmer from Deposit, the six directors. Six of the ten had served for 17 years or more; only Gifford and Higley had served for less than five years.

That this board was indeed bent on selling the fairground is confirmed in a *Reporter* article dated August 3, 1956 which disclosed that "the City of New York has recently evinced some interest in purchase of the fairground as a site for a sewage disposal plant. It costs the city $50 a foot to construct the proposed trunk sewer line for Walton and the fairground is the nearest available site."

In any event, the Board of Education put the fairground proposal up for a vote on December 9, 1956. It was soundly defeated, 712 to 320.

The first thing that happened after the first of the year was that at the January meeting of the Delaware Valley Agricultural Society, Peter Chambers, who had been a director for nearly 40 years and the president for 25, believing his term of office was over, announced his retirement. It was an act designed to have interesting repercussions. C. O. Field was elected president; Wallace Rich was elected vice president; Paul G. Williams, secretary; E. B. Guild, treasurer; and Mrs. Paul Schermerhorn, assistant treasurer. The other directors were Mr. Ogden, Mr. Camp, Mr. Higley, Carlton Chambers, and Mr. Gifford. Subsequently Peter Chambers suffered a mild heart attack.

What happened next was described by Paul Williams a little later in the year in reply to comments apparently made by George Rich:

"After the site was turned down by the voters of the Walton School system as a site for the proposed school, our Society sent notice to all stockholders of the Delaware Valley Agricultural Society, asking for a vote as to whether or not we should sell the grounds to the City of New York, and at an election, the proposition was carried in favor of selling.

"At a later date, the directors of our society met. There were six of the 10-member board present, and all six approved the sale of the grounds to the city."[1]

In its February 8, 1957 issue, *The Walton Reporter* announced: "Owing to sale of the fairground to New York City for a sewage disposal plant, the present management of the Fair has indicated it will not attempt a resumption of the local agricultural exposition this year."

Translation: No More Fair! The stage had now been set for a battle royal.

As the news leaked out, indignation began to boil, and farmers and 4-H people closed ranks. One observer has estimated that 27 heated meetings were held before the end of that historic and unforgettable year.

Emerging from the wings were Harold Carley of Delhi, 4-H Club agent, and George Rich of Franklin, president of the Delaware County Extension Service Association, a man known for a "get things done" quality of leadership, as well as for his years of devotion to the 4-H and for his top-notch herd of purebred Jerseys. They were joined by Milton Thomson, president of the Walton Central School Board, who also was a 4-H leader; Pearl Johnson of Meridale; Raymond Dann of DeLancey, representing the 4-H Club Department of the Extension Service Association; Ralph Gould of Hobart and Robert Hall of Bovina, representing the Agricultural Department; and Mrs. Henry Liddle of DeLancey, representing the Home Demonstration Department. They met with the Extension agents in the 4-H Club office and prepared a list of organizations and representatives to be contacted in order to form a substantial group that would do everything it could to stop the sale and keep the Fair going.

Carley immediately pointed out that without a 1957 Fair, the $10,000 premium aid which New York State gave fairs each year would be lost—perhaps permanently. Delaware County, with more than $20 million in farm income annually, is "one of the most important agricultural counties in the state" and "if any county needs a fair, certainly Delaware County does," he asserted.

Paul Williams, who had been executive secretary of previous fairs, had conferred with the Extension Services board of directors about the time of the fair board's January meeting and discussed fair problems. Now he, Mr. Carley and Harry Main, a prominent horseman and former fair board member, took the situation to the annual meeting of the New York State Association of Agricultural Fairs in Albany and came up with plans for a crisis meeting to be held on February 18 at what was then the State Agricultural and Technical Institute in Delhi. James Carey of Albany, the agricultural department trouble shooter for fairs, would be there.

At that meeting, attended by some 130 persons, George Rich agreed to head an Advisory Committee, and he named 21 members who came from all over the county and represented a broad spectrum of breeders, farm groups, business interests, the 4-H, Future Farmers of America and county government.

They included Mrs. George Wilson, James Dickson and James Van Alstine, Delhi; Charles Holdredge, and Leslie Sanford, Margaretville; William Gifford, Deposit; William Van Buren, Hobart; Mrs. Lawrence B. DuMond, Kenneth Merchant and Milton Thomson, Walton; Frank Slawson, Franklin, chairman of the County Board of Supervisors; Mrs. Alton Benedict, East Meredith; Frank Lamport, Hobart, Mrs. Henry Liddle, Raymond Dann and Robert Tyler, DeLancey; Alex Rabeler, Sr., Bovina, and Ted Roodhoof, Oneonta.

Also, Wilbur Cleveland, Bloomville; David Orton, Sidney; Mrs. Cecil Francisco, Sidney Center; and Michael Johnson, Meredith.[2]

State Senator E. Ogden Bush and Assemblymen Edwyn E. Mason and Edwin C. Howell all sent their regrets.

At a subsequent organization meeting, Mr. Rich named Frank Lamport, Milton Thomson, Michael Johnson, Mrs. Cecil Francisco and Ralph Gould to a more manageable group empowered to meet with "all parties concerned" and report to the larger committee. Raymond Dann was acting as secretary.

In short order the Advisory Committee set the dates of August 20-24 for a 1957 Fair but the problems seemed insurmountable.

Yes, there would be a Fair, but under whose auspices, the board of the Delaware Valley Agricultural Society which was bent on selling the fairground, or the persistent and hopeful members of the new group? How would the indebtedness be cleared and money raised to make the grounds ready and provide premiums (since the state reimbursement would not come until after the legislature met in the spring?) And the fundamental question: What to do if the contract went through and the fairground was sold? Still, at this point, a fair schedule was being prepared, and the 4-H classes were mostly set up.

At the April meeting of the Advisory Committee, Ralph Higley, president of the First National Bank of Walton, who had been a fair director since 1953, presented a few statistics: The total indebtedness of the fair was $8,256.49, consisting of three notes, he said, plus $13,000 in outstanding stock and attorneys' and legal fees "plus or minus $1,000." Mr. Williams, who had agreed to be personally liable for the notes, told the group that after the fair was reorganized in the 20s, it made money and was able to pay off its mortgage. But then the grandstand blew down early in the 40s, followed by the cattle barns. He said he would "hate to see the fair dropped."

Over and over the point was made that the Fair is not a mere local event; it is important to the whole county. It goes hand in hand with the future of agriculture in Delaware County, and it encourages young people through their demonstrations and exhibits to remain on the farm.

The pivotal events were two joint meetings of the fair board and the Advisory Committee, first at the Townsend School on April 25 and then in the courtroom of the Walton Village Hall on May 18.

A result of the first discussion was that New York City would be asked to surrender its purchase agreement and turn the fairground back to the new committee. Then, by floating a mortgage and selling stock, the Advisory Committee hoped to take over the fairground and continue to hold a County Fair.

At the meeting it was revealed that in addition to the stockholders' woes and the deficit, an estimated $800 to $1,000 would be needed to make repairs to fences and buildings.

Paul Williams speculated that Richard H. Burke, secretary of the New York City Board of Water Supply, might be willing to tear up the sale contract.

There was a slight glitch.

The resignation of Peter Chambers as president earlier in the year might not be legitimate because someone had made a mistake in counting the years and Chambers' term was not over. Legally, he was still president, not C. O. Field who had signed, as president, the purchase agreement with the City. Might the agreement be invalid?

On May 18, the two groups discussed the Fair "sometimes bitterly" before a packed house. The main results were:

- The existing directors agreed to establish a subcommittee to operate a 1957 Fair with the Fair Advisory Committee but only if that Committee would deposit $2,000 as a guarantee against unpaid bills.
- Ralph Higley agreed to resign upon three conditions: If the sale to New York City went through, if the note owed to Paul Williams was paid, and if a bill owed to Paul F. Eaton, attorney for the fair board, was taken care of.
- Other fair board members indicated that they might resign, too.
- The possibility arose that if a fair were held, it might take place in Afton—Chenango County, or in Morris—Otsego County. And that would leave Delaware County without any fair at all!
- Validity of the sale contract was again questioned.

Both Mr. Williams and Mr. Burke asserted that the contract had been signed "in good faith" and Mr. Burke said that, as a representative of the city, he would help "in every feasible way" to hold a fair. But Mr. Burke apparently was the target of many questions such as: Why not use the Loudon property for the disposal plant? His reply: The Loudon property would not be practical from an engineering standpoint.

The Walton Reporter described Milton Thomson, president of the Walton Central School Board of Education, and Mr. Burke as "playing verbal volleyball" over the Loudon property versus the fairground. The article reported one speaker suggesting that "a few who live in Walton have got this thing going without the rest of the county knowing it," and that the fair board "did not want to run the Fair and did not want anybody else to run it either." In short, feelings ran extremely high.

One often-quoted comment, reportedly made to Harold Carley, was: "If the business men can't make the Fair go, how do a bunch of farmers think they can do it?"

There were, naturally, the usual letters to the newspaper, by turns taking one side or the other, since there were apparently some who were perfectly willing to trade the fairground for a sewage disposal system, and the newspaper itself did not escape criticism. At one point Raymond Dann, who was later elected to the fair board and was to give strong support to the Fair all through the 70s, chided the paper by writing: "Many European countries claim that money comes before spiritual values in the U.S.A. I wonder how they would feel if they knew that one of the United States' leading dairy counties had a lead county paper who felt it should crusade against its dairy leaders to sell out its county fair to save New York City a few dollars.

"Many of Delaware County's farmers have gotten their start on prize money from the fair, and their interest in farming by raising cattle to show at the fair."

In another letter, a teen-age delegate to the National 4-H Club Congress in Chicago, wrote "to ask your help because I feel that New York City Board of Water Supply can find a suitable location for their sewage disposal without depriving the youth of Delaware County of their annual Delaware County Fair."

Helen Schriver Zandt, who then lived in Rock Royal, described attending the 4-H Congress as "the highest honor that can be bestowed upon any 4-H Club member in New York State," and she added that "The annual Delaware County Fair is the goal of the younger members, and as they grow older, their goal is the state fair... without this fair, many of the younger members of our county will not have any goal for them to achieve until they are older and then it will probably be too late..."[3]

A week after the Village Hall meeting, the Advisory Committee rejected the fair board's proposal that outstanding debts be assumed by the Committee in addition to putting up a cash guarantee. It wanted the $2,000 guarantee deposited in escrow as a cushion against possible losses, and it did not wish to take on any existing indebtedness. Further, it believed that a mortgage should be sought to take care of Paul Eaton's legal bill.

While all this was going on, the Donald W. Gleason American Legion Post in Delhi had an idea: Why not move the Fair to Legion Park, the old Delhi fairground?

Oneonta Daily Star staff writer H. Glen Harper, shortly after the May 18 meeting, wrote that Harold Tietbohl, Commander of the Donald W. Gleason Post, American Legion, who was a member of the Advisory Committee; and Raymond Secor, a member of the Post, had suggested the use of Legion Park as a permanent site. They had appeared at a meeting at Delhi Tech with an aerial photo of the grounds and a proposition.

Mr. Tietbohl said the Legion was in the process of "moving trees and brush, and improving the half-mile track. We also have a grandstand which will seat 800 persons, and it is in good condition." He said he thought the Walton grandstand, which heavy snow had brought down some years previously, had seated 1,000, and that the existing bleachers might seat 500. He added: "Anything you would do in Delhi would not be a one-year proposition. We would like to see you here." Said Mr. Secor, "There has been a lot of talk about the Fair merging with other county fairs. I can't see why (it) needs to go anywhere. As one of the top-ranking dairy counties in the state, if we can't support our own fair, we'd better quit."

This solution, of course, would have sent the County Fair back to its 1841 birthplace.

At the end of June came an announcement:

- A 1957 Fair would be held in Walton.
- George Rich would be president of the operating committee, Graham Jackson would be vice president, John Elderkin, secretary, and Milton Thomson Treasurer.

The next issue of *The Walton Reporter* reflected a rising tide of confidence, exhilaration and enthusiasm. Tuesday, the opening day, would be Band Day. High school bands from all over the county would participate and a marching band contest would be held in connection with the Firemen's Parade in the evening.

Joie Chitwood, the popular auto thrill driver, would perform.

There would be a talent show, tractor skill and pulling contests, a 4-H parade and horse-pulling events.

Ten dancing Palomino horses would entertain. There would be a square dance act, and for the first time in years—fireworks!

Reithoffer's Rides would be there as usual, together with the Home Bureau, Grange and extension services exhibits, a wildlife exhibit and a flower display.

Furthermore, the farm machinery dealers, who had been absent for some years, had been persuaded to return. The cattle exhibits were larger, Grange exhibits came piling in, and a new feature came on the scene: The Talent Show.

All in all, the 1957 Fair burst into the records as a celebration instead of a disaster with the attendance on opening day topping 7,000, which made for a tidy gate receipt even though all of the firemen were allowed in free!

As the dust settled and Milt Thomson began to tote up the income and expense sheets, it became plain that the Fair had made a profit.

The $2,500 which the County Board of Supervisors had placed in escrow for the Fair had not been touched and would be returned. Premium awards had cost almost exactly the $10,000 that New York State provided, so there was no loss there.

The final tally in December showed total receipts of $23,912, plus $10,000 in state aid, or $33,912. Expenses, including $11,568 paid in premiums, were $23,491, leaving a "return above costs" of $10,421. After paying $4,000 on the outstanding note and returning $2,500 to the County Board of Supervisors, the profit for 1957 was $3,921.

The battle had been won and the Fair was set to survive.

But what about that contract, duly signed by both parties, handing over the fairground to the City of New York for a sewage disposal plant? And what about those notes held by the First National Bank in Walton?

Subsequently, the City did, indeed, build a sewage plant on the Merritt Loudon property downstream from the fairground. How come?

The answer is that, yes, the battle was won, but the war was not over.

In an editorial on December 13, *The Reporter* deplored the conflict between the need for a "free sewer system" on one hand, and the need for a County Fair on the other and commented crossly: "If the same amount of thought had been put on finding a way of obtaining both these objectives as has been spent in scrapping, something might have been accomplished."

Meanwhile, several things were happening:

- Delaware County Judge Walter L. Terry in November granted a petition presented by the "old fair group" asking that they be given a court order permitting the sale of the fairground.
- But on December 5, Judge Terry vacated this grant of permit saying that the membership of the Delaware Valley Agricultural Society should have been notified. At the same time he called for appointment of a referee.
- Richard H. Burke, secretary of the New York Board of Water Supply, announced that the city was ready to close the deal.
- The group of farmers who had united behind George Rich, Harold Carley, Milt Thomson, John Elderkin, Graham Jackson, and others, began to accumulate stock in the Fair organization for what was to become a bitter stock and proxy fight. If the "farmers" could remove the entrenched officers, they could probably settle the problem of the sale contract.
- As far as the notes were concerned, it paid to have friends. Henry Eckert and Clark Hillis of the First National Bank of Hamden were willing to lend the funds needed to pay the fair's debts.

The farmers also had another ace up their sleeves. They called upon Gleason B. Speenburgh, an attorney from Margaretville, who was the Delaware County District Attorney, to help them.

It was the annual meeting of the Delaware Valley Agricultural Society on December 21, 1957 that finally decided the outcome.

Appearing at that meeting, Speenburgh tossed in challenges and objections as the voting for officers got under way.

There were heated exchanges over the right of holders of 1/5 shares to vote, over the appointment of tellers, and over the right to vote by proxy.

The meeting, with 56 present, had an unprecedented attendance, but the farmers' group lost the election. The "old" directors were returned to office: Ralph L. Higley was elected president; Mrs. Paul Schermerhorn, secretary; Paul G. Williams, treasurer; and Louis F. Camp, Sr. and Bruce Curry, directors.

At the end, Mr. Speenburgh described the whole scene as a "goofed up" election. It was obvious that he was not calling it quits.

However, the election was not the most important thing that happened at that annual meeting.

Raymond Dann of the farmers' group made a motion that the Delaware Valley Agricultural Association go on record against sale of the fairground to New York City. It passed, 34 to 0.

This motion was forwarded to Judge Terry by George Rich.

Early in January, 1958, Judge Terry issued a decision voiding the sale of the fairground to New York City, saying that he felt he would have to do so on the basis of that vote.

Almost immediately seven of the "old" directors agreed to resign if the "new" directors would agree to settle the unpaid bills and soon-to-become-due $8,000 note held by the Walton bank.

Again Gleason Speenburgh was on hand to guide his clients through the talks and financial complications of this long and complicated affair.

In the end, the seven directors resigned: Ralph Higley, Paul Williams, Louis Camp, Grace Schermerhorn, Wallace Rich, C. O. Field and Bruce Curry. Three others, Carlton Chambers, Henry S. Ogden and William Gifford, went on to serve with the new organization.

Elected as officers were the men who had organized the successful 1957 fair:

George Rich, president
Graham Jackson, vice president
John Elderkin, secretary
Milton Thomson, treasurer

Directors, in addition to Mr. Chambers, Mr. Ogden and Mr. Gifford, were Steven Vamosy of Stamford and Vanley Blackman and Herbert Elderkin, both of Walton.

The 1958 Fair was successful and profitable also. It marked the beginning of a new era, years of growth, improvements and innovations which have made the County Fair at Walton one of the finest in New York State.

Walton Historical Society Photo

The Fair in the mid-30s still used the old wooden cattle barns, foreground, and the grandstand had been expanded with bleachers. It was destroyed during a windstorm during the 40s. Parking was inside the track. In the very center was a baseball diamond.

Photo courtesy of Iris Elderkin

By 1960, bleachers, with a tent over the top, were serving as a grandstand. The horse show ring was still opposite the grandstand, but the baseball diamond was gone. Tents for the cattle and other livestock had grown longer, and cars were beginning to be parked on the Loudon tract, later the Ellen Hanford Memorial Field. There was still harness racing, but it disappeared in 1964.

Photo courtesy of Veda Thomson

By 1976, a new grandstand had been erected, with a stage opposite, and the horse show ring had been moved to its present location. Cattle were still under tents, but there was a new 4-H building.

Photo courtesy of Veda Thomson

The three big new cattle barns, built in the early 80s, take the spotlight in this photograph.

Walton Historical Society Photo

By the time the Centennial year arrived, there was a new horse barn, center foreground, a practice space and holding area for the horse show, a new draft horse barn, a 4-H dormitory—and the long concession tents were in place.

Aerial Photo by Robinson

After the January 1996 flood, the fairground was a mess.

Aerial photo by Robinson

The Delaware County Fair at Walton, 2000

Approaching a Centennial

Aerial photo by Jim Robinson, courtesy of Veda Thomson

The fairground in 1983.

After the crisis of 1957 was past and there was no longer any question of whether the Fair would succumb to the efforts of New York City to build a sewage plant on its fairground, there was a long period of renewed optimism and activity. Growth and expansion were the watchwords as the Delaware Valley Agricultural Society marched towards its Centennial—100 years since its beginnings in 1887.

To prepare for the 1958 Fair, the new board took much the same attitude as a present-day CEO faced with Chapter 11—reorganize, retrench, but keep innovation and customer satisfaction at the top of the list.

The board decided that hereafter the directors would not be paid. And there would be no free passes—for anybody—the directors agreed to buy their own tickets. They cut three classes from the horse show and pared down some premiums and compensation, thus finding money needed for the harness racing. Ticket prices were raised slightly. But instead of waiting to pay premium winners until the State funds were in hand, the directors decided to borrow the money, pay promptly, and pay off the note when the funds arrived in the spring.

They spent $1,700 on capital improvements including an addition to the Grange Building, painting, and fence repairs. Herbert Elderkin, grounds chairman, was predicting a much spruced-up fairground. The board voted to broaden advertising coverage and occasionally play music over the sound system. Radio coverage of the Fair was discussed. The directors started to explore using the Loudon field for parking. All of these things were in tune with a new-era atmosphere that made preparations for the six-day event, August 18-23, 1958, a stimulating and challenging time. The Fair would run a day longer than usual.

Cattle would arrive on Monday and remain until Saturday; there were several new classes. Ralph Taylor announced that there would be a "cattlemen's banquet," and with that, a long-lasting tradition began; the banquet is still a highlight of the cattle department's week at the Fair.

At the horse show, Mrs. George Wilson of Delhi and her committee were planning a grand parade headed by a king and queen.

Vanley Blackman announced some "refinements" in the tractor pulling; Bill Tweedie forecast "bigger and better" Grange exhibits.

Tuesday, Firemen's Day, would feature not just a parade but all kinds of exhibits—fire-fighting equipment, fire prevention information and a first-aid contest. Leslie Sanford planned special effects for the department parade.

Millard Kilpatrick declared the revival of harness racing well in hand. The Fair had now become a member of the U. S. Trotting Association.

The perennially popular Hell Drivers would appear on Monday.

Bob Howe would play the organ, there would be a junior Talent Show, and on Friday and Saturday nights—a rodeo.

Also, for the first time, a six-horse hitch would come to the Fair. The big, feather-footed Clydesdales had previously appeared at the State Fair and would be sent by the Metropolitan Bargaining Agency, a milk cooperative. And to cap everything else, there would be fireworks.

When the week was over, tired but happy exhibitors, visitors and board members declared the 1958 Fair a winner. Furthermore, it had made a profit of $2,500.

That is not to say, however, that the whole week went smoothly.

There was a terrific storm which knocked a seat off the ferris wheel and threatened to pull down the Clydesdales' tent.

A team of heavy draft horses "bolted as if shot from slingshots" as the horse-pulling contest was getting under way and made for the crowd at the fence. The people scattered, all but two women who seemed to be frozen to the spot, but the horses separated and dashed to either side of a snow fence in front of the race horse barn.

At the cattle show, a youngster was butted by a young bull as he was pinning on its blue ribbon. The boy was knocked over the stall and into a feed bin. By 1984, senior and two-year-old bulls had been taken out of the fair book and not allowed on the grounds.

A young boy happened to pick up some unexploded fireworks and was taken to the hospital. Another boy was injured when a horse fell on him.

Cited as a hero was 11-year-old Dale Tompkins of Liberty, who, astride a pony, was able to halt a runaway pony cart.

The board minutes for 1959 show a concern, in the spring, for a better supply of water and later a need for more space in the cattle barn. The importance of keeping the quarter-poles on the race track clear was stressed. The area inside the track was beginning to build up.

In October, the directors voted to ask the County Board of Supervisors for $5,000 to match an equal sum the Agricultural Society would spend to build a 50-by-196-foot cattle barn. Toward the end of the year, the board began to think about acquiring additional land, and it cast its eyes upon the Merritt Loudon flat.

After the 1959 Fair, State Inspector James Carey of Albany is quoted as saying: "It is my belief that Delaware County has a real good county fair—its cattle show is extensive and excellent. The horse show creates a lot of interest, especially the classes for youngsters. The 4-H show is outstanding in every department. Home demonstration is interesting. Grange exhibits are both educational and outstanding. I am also impressed by the large display of farm machinery.

"People couldn't ask for a cleaner and more active midway. I am greatly impressed with the spirit exemplified by the officers and directors who devote much time and effort."

In 1960, a float parade, which had been so much a part of the early fairs, was restored. Once again it became a popular contribution to the festivities and it continued until the early 70s. Individuals and organizations, industries and small businesses were eligible to compete for prizes ranging from $10 to $40. *Industries of Delaware County, Past and Present* was the initial theme, followed by others such as *Historical Monuments, Famous People of the United States, Events in the Life of a President*, and *A Day to Remember in School.*

The 1960 Fair featured an Ice Show, complete with organ, scenery and a fancy-costumed troupe. The Jack Kochman Thrill Show returned once again. Reithoffer's Rides gave away two bicycles. Ferris Terry volunteered to supervise a horseshoe pitching contest, an activity popular for many years but, late in the 1980s, it gradually faded away.

At this time, the horse show ring was opposite the grandstand with the show ring bleachers facing the track. Looking across the track from the grandstand, a 1963 map shows the tall announcer's booth for the horse racing and next to it the stage and dressing rooms. The WDLA trailer was behind these structures. The secretary's office was on the upper edge of the stage.

Concession tents were spread around the area, but they were few in comparison to today. There are only 29 numbered items on the map. The cattle tents were approximately where the new barns are now, with two horse barns beside them, and there were two more horse barns next to the grandstand. All of the animals in the horse show had to cross the track. The ticket office was on Fair Street with a second gate on More Avenue. The men's and women's bathrooms, widely separated, were way behind the Midway. Where the Post Office is now, was a milk bar where fairgoers could get sandwiches.

The announcer's booth, opposite the grandstand, was a more or less sacrosanct place: Only officials were allowed. Its guardian was a well-known Walton personality, Snowball, a very black man with a shock of white hair, who has been variously described as "a father figure," "a very good guy" and "a godsend to the Fair."

He had arrived originally with a circus which came to town, and when it left, he stayed. Bill Austin gave him a place above the sales stables, where he also worked, and he did many chores around town including the spreading of sawdust for the weekly summer Delaware Street square dances. He was an unforgettable personality in the Walton community, a friend to many and many were his friends, including lots of youngsters. One was the current fair board president, John Jackson. Another was Walter Rich, son of the then Fair president, George Rich.

Both men have warm memories of Snowball—especially in view of the fact that on occasion he allowed them to climb up into the holy-of-holies, the announcer's booth, from which perch they could watch the stage show.

As it turned out, Walter was more interested in trains than fairs; he would sit beside his grandfather with a stop watch and clock the races, but he says he also spent a lot of time sitting on the river bank watching the trains. He had a booth at the Fair containing an O and W Railroad exhibit—a place he preferred to the dairy barn. But both Walter and his sister, Margaret, showed livestock and were active in the 4-H. He grew up to be the Chairman of the Board, President and Chief Executive Officer of the New York, Susquehanna and Western Railroad with offices in Cooperstown.

A new 4-H building was constructed in 1963 at a cost of $13,000. Ward MacMillen, the new 4-H agent, praised the directors as "the most cooperative fair board I have worked with," and pointed to the more than $21,000 worth of permanent improvements made since 1959, including cattle stalls, toilets and the new building. There were 3,000 exhibits in the junior department that year. Milt Thomson reported that the Fair made "slightly more than $3,000" in 1963.

As 1964 made its entrance, an old-time favorite at the Fair, horse racing, made its exit. According to some long associated with the Fair, horse racing was discontinued for at least two reasons: Times were changing and the State Association ruled it out.

The first thing the 1887 founders had done was to lay out a half-mile track for the all-important horse racing.

Through the years, however, all of the growth and expansion took place inside of the track—all of the buildings, tents and displays, roads and exhibition rings. For the first 50—even 75—years this was fine because the Fair was small, but then it mushroomed. More and more buildings were erected until at last the racing judges were unable to see the quarter posts for all the obstructions.

Many were sorry to see the horse racing go, none more than fair board president George Rich, whose love of horses was part of his life. That was the only time this usually self-contained man was obviously upset, recalls his son, Walter. "And he was *very* upset."

Horse racing has survived at several area fairs, but it has never regained its old-time appeal.

A contest for the best-decorated dairy barn stall was initiated in 1964, and in 1977 the contest was given a theme: *The Story of Our Farm*. Subsequent themes have included *I Am Proud to Be a Farmer, Farming Is Udderly Wonderful, Promote Dairying Through Advertising, Farming Is Our Business, Farming Through the Years, 101 Ways to Farm, Welcome to the Real World, Farming is a Family Affair, The Tools of Our Trade, Families Make Farming Fun, Champion Cows* and *Champion Kids*.

The late 60s produced more building: A new grandstand in 1965 for $16,822, a $6,000 milking parlor in 1967, and in 1969, a new entrance gate and sign. There were also new programs such as an open class for beef cattle, the cattle stall decoration contest, and the demolition derby.

In May, 1968, the fair board spent $41,000 to buy 25 acres of land from Merritt Loudon, whose abutting field it had been renting for some time as a parking area. Later, a portion was sold to the Village of Walton. The Loudon field was purchased with the aid of a trust fund called "The Platt Mead Hanford - Jennie Bradley Hanford Memorial Fund," better known more informally as The Ellen Hanford Fund.

Ellen Hanford, a descendent of early settlers, was born in Walton, December 23, 1900, attended Walton schools, and was a member of the Congregational Church. She earned degrees from Cornell and New York University, did graduate work at Columbia and the New York State University at Albany, and spent the rest of her life as a teacher, first in Walton, from 1923 to 1925, and then in Herkimer and Utica.

At the time of her death, in March, 1963, Miss Hanford was teaching English at the Utica Free Academy, but she had not forgotten her native town. She left a will setting up the memorial trust fund which provided for income to be paid to several non-profit organizations including the Delaware Valley Agricultural Society, the Walton Historical Society, the 4-H and the William B. Ogden Free Library. Many improvements at the fair have been made possible because of Ellen Hanford's bequest, and the Loudon property was subsequently dedicated as The Ellen Hanford Memorial Field. Green and white markers at the entrance to the field pay tribute to Ellen Hanford, "a gracious Walton lady" whose "generosity has provided funds for us to continue ongoing improvement programs."

The Fair had been growing steadily through the 13 years following its reorganization by the committee of "farmers" who had rescued it from sale to the City of New York in 1957, but during the 70s and 80s it zoomed.

The expansion project in 1972 produced a $10,000 youth dormitory and in 1973, $3,672 worth of work on the stage. A new roof on the grandstand in 1976 cost $16,685 and the huge dairy cattle barn complex in 1979 cost approximately $79,000. Of this, donations accounted for $39,000 including a $5,000 check from an anonymous donor. "*Somebody* loves the fair," one of the directors observed. Primo Construction Company was the builder.

In their 1975 report to the board, fair committees reported 407 dairy cattle and 83 beef, something of a record. There were 37 swine, 91 tractors, 25 or 30 draft horses, and 100 smashed-up cars in the Derby. The Society paid out $18,210 in premiums, another record, and it had "only $400 left on the original bank mortgage of $10,000." The Society was, the directors said, "in a good financial position."

But apparently 1977 was even bigger. Saturday of Fair Week "was probably our biggest day ever," fair officials said after they had sold 4,585 single-admission tickets. A total of 17,068 single admissions had been sold for the six days, plus 848 season passes and more than 2,000 exhibitor cards, bringing total attendance to approximately 20,000.

In 1976 the 5 1/2 x 8 1/2 Premium Books began to sport two-color covers and for some years they even had color-coded pages.

As preparations for the 1978 Fair were under way, the Rev. Arthur Lefgrin, a retired clergyman from East Meredith, approached the board about the possibility of having a Sunday evening church service "for the convenience of those working at the fairground, setting up exhibits and caring for animals, making it difficult for them to attend Sunday services."

The Rev. Mr. Lefgrin officiated at the first service, assisted by two men from Walton. Arthur Jamieson, in charge of a brass ensemble, was a music teacher in the Walton school system. Eugene Wieand, superintendent of schools, directed an ecumenical choir. Mr. Lefgrin's idea became an annual happening which grew as the years went by.

On August 13, the Sunday before the Fair in 2000, at 7 PM, the 23rd service of this kind took place. "The Church in the Wildwood" was the subject of the speaker, the Rev. Tom Clemow, Jr., district superintendent of the Oneonta District, Wyoming Conference, the United Methodist Church. Mark McCarty, band director at the Downsville Central School, and senior choir director at the United Presbyterian Church in Walton, led a large choir made up of participants from all over the county. The organist was Jody Jamieson of the United Presbyterian Church.

To round out the decade, in 1979 three stone piers were erected at the entrance to the Ellen Hanford Memorial Field.

One of the first events of the 1980s was the beginning of the Fair Book Cover Contest in the schools. Early in 1982, Marcia Kozebec of Walton, who was with the Delaware County Council on the Arts, discussed the idea with George Rich who agreed to get it organized. The fair board offered $15 for the first-prize drawing, which would go on the cover; $10 for second and $5 for third, both of which also would be used. Hundreds of county children have participated in this project since 1982 and the premium books, which have grown in size to 8 1/2 x 11, now contain 20 or more contest drawings.

The need to expand continued. Lighting poles were installed in the parking field and in Trailer City, which was growing each year. The Midway needed more electricity. The trophy stand at the horse show ring needed fixing, and a holding ring below the existing horse show ring was put into the works. The foundation of the Grange Building had to be repaired. Fifty feet was added to the grandstand, and the subject of new rest rooms or improvements to rest rooms were constantly in the minutes of the fair board meetings. The big projects for the 80s, however, were the Stan Budine cattle barn in 1981, the Paul Mattern Livestock Barn in the spring of 1984 and the draft horse barn in 1986.

Stanley Budine, Jr., who died in April, 1981, had been a member of the fair board, and he was also a director of the Delaware County Holstein Club. He was a former president of the county Farm Bureau. The 1981 Fair Premium Book was dedicated to him for his "friendship, his dedication, and his adherence to the goals of the fair...a fine man and a fine farmer."

Paul Mattern, who died in 1983, had been an Extension Agent in Delaware County for two decades and was active in dairy and livestock affairs at the Delaware County Fair for many years. He had received the Distinguished Service Award of the State Association of County Agents in 1974. The 1983 Fair Premium Book paid tribute to him and noted: *"Probably no one helped bring more improvements to farming in Delaware County over the past 20 years than Paul."*

Bids for the Paul Mattern barn were received in February, 1983 with Warren Peck low bidder. Even before the project was under way, the Jolly Barnyard 4-H, under the leadership of Jean and Don Tweedie, raised $1,120. Again, the anonymous donor, who had given $5,000 toward the barn complex in 1979, came forward with a like amount. Later, to help finance the project, donor poles were sold at $50 apiece. The final cost, excluding bleachers, benches and wiring, was approximately $56,000 with nearly $27,600 in contributions.

There were concerns during these years other than physical improvements and construction projects, however. There was a point at which the then Gov. Mario Cuomo was proposing to cut state aid to county fairs out of his budget. The New York State Association of Agricultural Fairs put a lobbyist to work, and George Rich, Gary Tweedie and Arnold Froehlich went to Albany. They reported the trip "beneficial."

However, a new threat was coming: a real estate tax. To counteract this and anything similar, the board decided to change the by-laws and make the Society a not-for-profit membership organization rather than a stock company. Carl Mugglin guided the board through this procedure.

In the beginning, the Fair had been formed around a membership concept. In 1887, tickets were sold which entitled members to enter the fairground and exhibit items. However, when the financial problems of the 1920s reared their ugly heads, stock was sold to help ward off disaster. This stock and its value to stockholders was one of the issues during the crisis year of 1957. After the reorganization, the Society continued to be a stock company.

In the 1980s, however, under the not-for-profit corporation law, the Fair could not legally hold stock, and a change was advisable. Carl Mugglin suggested two other modifications:

- Create one additional directorship, giving the board 11 members (and thus have a tie-breaker).
- Specify that a candidate at board elections would run against all other candidates; those receiving the highest number of votes would be elected.

At a reconvened annual meeting on June 18, 1985 the membership approved the by-law revisions.

Along with these issues was the continuing one of insurance. Well into the Fair's history, liability insurance were words hardly ever mentioned in the minutes, but by the 1980s they were there constantly. Claims were mounting all over the country, premiums were skyrocketing; at one point an insurance representative appeared before the board and suggested writing to legislators and trying to "educate the public" about the folly of awarding large settlements. Finally, like a virus, the epidemic of ballooning liability premiums spread to the concessionaires and to entertainers. All now had to prove insurance dependability.

Back in 1924, when the Fair's existence was first threatened, there was a great deal of public sentiment for using the fairground as a gathering place for community events. After the grounds were sold at auction and the Fair continued, little was heard of this idea, but by the time the 1980s had arrived, the earlier dreamers would have pranced for joy. Board minutes of this period show that so many groups and organizations were asking permission to use the fairground at times other than during fair week that the directors were hard put to grant all the requests.

People wanted to hold horse shows, flea markets, pet shows, art and craft shows, and cattle shows. The Soil Conservation office wanted to give out trees, the Fire Department sponsored a Winter

Weekend, the Booster Club, JCs, Scouts and National Guard all asked for time and space and the 4-H events were many.

From the beginning, the Fair had had a reputation for avoiding the more rowdy elements, so when there began to be a problem with alcohol in the early 80s, it did not last long. Board members voted to "take whatever steps (are) necessary to reduce the use of alcoholic beverages on the fairgrounds." They put up permanent *No Alcoholic Beverages* signs, and by seeking the assistance of the sheriff's department and the local police, they made it abundantly clear that the edict would be enforced. The food concessions also were managed with a tight rein. The board insisted upon approving whatever food was to be sold. In 1985, the county Health Department commended the Delaware County Fair "for superior conditions."

The 80s also accounted for three anniversary celebrations: Reithoffer's Rides observed its 50th year at the Fair in 1984, the Town of Walton celebrated its Centennial in 1985, and the Fair itself became 100 years old in 1986.

The better part of a year was spent getting ready for the Centennial. The last in the long series of building and maintenance projects was authorized early in 1986, including a new draft horse barn. As the months passed there were myriads of program decisions to be made.

C. B. Bassett and his comrades of 1887 would not have recognized the fairground. It was no longer a fairly empty, flat meadow with a few barns and tents scattered about. What had arisen looked like a small city, a good part of it developed during the 30-year tenure of George Rich and company.

There was the long, H-shaped cattle barns, new horse barns, a new horse show ring with a practice space, holding area and covered bleachers; a new 4-H building and youth dormitory, an FFA building, an enlarged grandstand with subsequent additions, the new Paul Mattern livestock barn and the Stan Budine barn, a new stage with a roof over it, a splendid new concessions complex, lots of rest room facilities, more roads and an area for trailers, caravans and campers that was a village in itself.

One observer estimated that there were still nearly 90 tents, but there had been a serious effort to cut down the number. At one point, there had been a tent over the grandstand, and "putting it up was a pure pain in the neck." With the increased facilities on the grounds and lots of space in the Ellen Hanford Memorial Field, there was no difficulty accommodating cars for the record 7,340 visitors who came on the last day. The final attendance tally was 22,587. What was most extraordinary—the sun shone every day of the week!

People came from far and near to be part of the festivities and meet old friends. One of these was ninety-one-year-old Jay Cook who had been born on a West Brook farm and started to show sheep at the Fair when he was twelve. For 60 years, he showed various kinds, including his favorites, the soft-fleeced Merinos. He was evidently impressed by the Fair, "much larger and better than I expected," and he commented on the large number of hogs. "We don't have anywhere as many animals at our fair (near Trumansburg) any more."

Highlight of the week, of course, was the Centennial Parade, *Memories of the Good Old Days.*

An estimated 1,660 persons watched on Thursday night as the procession of colorful floats, antique cars and tractors, fire equipment, horse-drawn carts and carriages and horseback riders came down the track, led by the Old Tyme Band of Bainbridge. Prizes, totaling $3,000, were awarded.

Winner of the prettiest float was the Walton Central School Class of 1965. Its picturesque creation, entitled "The Good Old Summertime," featured a group in 19th-century costumes standing on a bridge. Beside them, on a bandstand under a gazebo, musicians played the theme song. Lorraine Bryden directed the band; others on the float included Walter, Donna and Kendra Smith, Marie Ray, Charles Goodrich, Erin, Kristin and Janice Ewain, Robert Miller, at whose place the gazebo and bridge had been built; Kari Smith, Mary Maurey, Tim Gladstone, Janice Race, Connie Gage, Jennifer Miller, Sean McCoy, Terry Rogers, Kelly McCoy, Eleanor Oles and Nadine Kopetz.

Judged the most original float was that of Delaware County Porkies Plus of Sidney Center, and the funniest, Kraft Dairy Group.

After the parade—Fireworks!

The Centennial ran for six days, August 11 - 14. In addition to the float parade, the six-day horse show, the cattle, sheep, swine and goat shows, the art show, farmers' market and 4-H and FFA demonstrations, the schedule included a Bluegrass Festival. There was a firefighters' parade, a talent show, a tug-of-war, tractor and truck pulls and for two nights, the Demolition Derby. Reithoffer's Rides observed its 52nd year at the Fair. On the grounds was what was described as "the largest exhibit of farm machinery ever."

And then there was the food. Asked what she liked best about the Fair, one woman said: "You can get a baby blanket for $10, and you can eat everything from hot dogs and pizzas to shrimp and grilled chicken, with or without french fries."

George Rich, who completed 30 years as fair board president in 1986, remarked in an interview that the Fair was successful through the years because the people of the county worked hard to make it happen. "We've had the cooperation of the whole county," he said. "That's the only way you can operate."

During his tenure, Mr. Rich said, exhibits in all departments had increased tremendously. He said the Fair at its Centennial had one of the largest cattle exhibits of any county fair in the state. As far as the extensive building program was concerned, Mr. Rich said: "One of the things I'm proudest of is that they are all paid for!"He praised the dedication of board members. Milt Thomson, the treasurer, was the only member to have served as long as Mr. Rich. The others, all of more recent vintage, were Douglas Bull, Robert Cole, Arnold Froehlich, Richard LaTourette, John Jackson, Jon Seeley, James Taggart, Don Tweedie and Gary Tweedie. Of these, John Jackson, Richard LaTourette, Jon Seeley and Gary Tweedie are still serving in 2001.

Photo by Robert N. Lennek

In a dairy county, the cattle barn is the central point at the Fair.

Photo by Robert N. Lennek

Many honors at the State Fair in Syracuse have been won by 4-H members from Delaware County.

Photo by The Delaware County Times

Throughout its history, the tractor pulls have been one of the most anticipated events.

Photo by The Delaware County Times

No Fair would have been complete without a Tug-of-War in the early days, but then they faded away for a few years. Now, as the Centennial approached, they came back to a newly appreciative audience.

Walton Historical Society Photo

Climax of the Centennial Celebration in 1986 was the float parade featuring horse-drawn carts and carriages, antique cars and tractors, fire equipment, horseback riders, and, of course, colorful floats. Declared the prettiest float was the one created by the Walton Central School Class of 1965.

Walton Historical Society Photo

It's hard to realize the tremendous contrast between 1910 and 1986 until you see the actual line of march. Here are the wagons, buggies and horses of an earlier time.

From the Cornell Cooperative Extension Archives

To commemorate the Centennial, a 4-H alumni reunion was held on the stage.

From the Cornell Cooperative Extension Archives

There was also a four-tiered anniversary cake with lots of memory pictures to look at.

Walton Historical Society Photo

Then. . .

Photo by The Delaware County Times

And now! The intrepid driver is Gay Cochran.

Photo by The Delaware County Times

Some things at the Fair are timeless, such as an exploratory walk past the food concessions. Pizza, sausage with onions and peppers, hot dogs, grilled chicken, French fries, chili, spaghetti, ice cream—you name it, it's there!

From the Cornell Cooperative Extension Archives

When the 4-H horse barn was built in 1976, 4-H members helped paint the outside.

From the Cornell Cooperative Extension Archives

Paul Mattern, an agricultural agent in Delaware County for 20 years, was a hardworking presence at the Fair and much esteemed for his efforts to improve farming in the county. One of the big barns built in the early 80s was named for him. He is seen here with Kathy Johnson at a TV "apple pie" demonstration.

Photo by Robert N. Lennek

The idea is to start when you are very young.

Photo courtesy of Irma and Leo Pierce

It isn't every day. . . that you can rest your back. . . on a bale of hay! There have been many Tweedies from the very beginning of the Fair. Here is Harold, of Mundale, taking a breather in the dairy barn, 1980.

Photo by The Delaware County Times

The Firemen's Parade, traditionally held on Tuesday night, has been part of the Fair as long as anyone can remember.

Reaching the Millennium

Photo by The Delaware County Times

One hundred sixty years after the state legislature passed its "Act for the Encouragement of Agriculture," the Delaware County Fair is widely regarded as one of the best in the entire state because it has made agriculture the centerpiece of its existence. Now, in the 21st century, as people close to the Fair look back to 1841, when county farmers gathered their animals and their vegetables at Judson's Inn in Delhi, and to the Summer of 1887, when C. B. Bassett and his friends formed the Delaware Valley Agricultural Society, there is reason to be proud.

After the excitement of the Centennial, the fair board more or less rested on its laurels for a few years, but it did continue a long-range maintenance and building program. New construction at the end of the 80s included a rest room near the Grange Building, a 500-foot addition to the grandstand, a chain link fence around the stage area, and a larger, L-shaped commercial tent.

George Rich, returning from the annual New York State Fairs Association meeting, at the 1989 fair board luncheon reported that the Delaware County Fair was apparently doing things right: "Their advice was to get people involved and not try to do everything yourself," he said. "And this is what makes our fair so successful—so many people are willing to get involved and help."

The 1989 Fair went on as scheduled, and at the annual meeting in January, 1990, it was announced that the Ellen Hanford fund would provide $6,300 toward the new horse barn. With Don and

Gary Tweedie re-elected, board membership remained unchanged. After the election of officers in January, 1990, the board consisted of, once again, George Rich, president; Don Tweedie, vice president; Arnold Froehlich, secretary, Milton Thomson, treasurer; and, Doug Bull, Max Dutcher, John Jackson, Richard LaTourette, Jon Seeley, Jim Taggart and Gary Tweedie.

On May 23, George Rich died. He was 74 and had spent most of his adult life in leadership roles at local, county and state levels. The entire county grieved.

John Jackson says he was driving along the road with his radio on when the news came over the air. Stunned, he called Don Tweedie who would be the new president, and soon events were set in motion to carry on. Tweedie was to serve for the next 10 years.

These were years during which agriculture in Delaware County continued to decline, but shortly before he retired in 2000, Tweedie told a news reporter: "If you came to our Fair, you wouldn't think so." He said, "This Fair attracts more than a thousand head of animals, and it's the largest event that happens in Delaware County the whole year." It started as an agricultural fair, and it has consistently tried to remain that way. More than 70,000 persons attended the Fair in 1999. Still, as Tweedie pointed out, "keeping the focus on farming has meant turning $30,000 to $40,000 a year back into agricultural exhibit space and supplies."[1] Furthermore, one of the best attractions seems to be top-grade—and therefore expensive—entertainment, like Montgomery Gentry, Mark Chestnutt, Jessica Andrews, Jo Dee Messina or the old favorite, Joie Chitwood. (Actually, there have been several Joie Chitwoods. There was Joie Sr., who was the daredevil of the 50s, then there was Joie Jr., and Tim. The act went out of business after the 1999 season.)

Fairs have become big business. The annual construction costs, entertainment fees, periodical increases in exhibitor premiums, insurance costs and contingencies are an ongoing challenge. In 1999, nearly $50,000 was spent for top-row entertainment and more than $80,000 for various kinds of permanent improvements.

Major projects during the 90s included doubling the size of the grandstand and a new roof, many more rest rooms at more convenient locations, a new Fair Office and Museum/Store, a Post Office offering a special Fair cancellation, extensive plantings to beautify the grounds, largely contributed by a local nurseryman, Vern Wildenstein; improvements to the 4-H dormitory, complete resurfacing of the horse show ring, extensive repairs to the track, and concrete floors for the sheep, goat and pig barns and new pens. Still, in 1999, $50,000 went into agriculture, "the number one priority."

Midway through the decade Mother Nature threw Delaware County—and the Fair—a curve. On Friday, January 19, 1996 a rain and sudden thaw melted the winter's four-foot snowfall so fast that the torrents rushing down the hillsides caused devastation, many said, equal to or surpassing the flood of 1935. Once again, Delaware Street in Walton was inundated, and, curiously, once again, during the flood, a fire destroyed local businesses, this time, the Family Shopping Center and Charlie Fiumera's hardware store. All over the village, and in fact throughout the county, the damage was so severe that a state of disaster was declared; it took many months for much of the damage to be repaired. At the Fair, board members looked ruefully at the washed-out track and deep pits everywhere and wondered if a 1996 Fair would be possible.

But as the consternation and dismay subsided, the entire community presented a united front to restore the fairground. The horse show people, the Boy Scouts and Cub Scouts, the 4-H, the FFA, the BOCES student body, various fire departments, and even the National Guard with its heavy equipment, pitched in. Members of the fair board worked, sometimes almost around the clock, in an exhausting effort to have the grounds in shape for the August deadline. Donations poured in from many sources, and individual volunteers gave many hours of time and labor. The Walton Grange held a benefit auction, raising nearly $800, and the Hamden Senior Citizens Group, the American Red Cross and the Allegheny County Fair each sent $500. In all, donations totaled $16,000 to help offset the flood damage costs of $72,000. But that figure does not begin to reflect the amount of human effort. Funds were requested, but were not forthcoming from (FEMA) the Federal Emergency Management Administration. When the 110th County Fair opened on August 12, 1996, hardly a trace of the damage remained. On the inside cover of the Premium Book, however, were these words:

The Board of Directors of the Delaware Valley Agricultural Society dedicates this year's 1996 Fairbook to our friends who donated gifts, their time and efforts to restore the January 19 flood-ravaged fairground to the beautiful condition you see today.

The following year marked the death of Milton Thomson, another of those persons so highly involved in the crisis of 1957. He had served as fair board secretary for 37 years, until his retirement in 1994. His wife, Veda, created a scholarship fund in his memory.

The end of the Walter Shackelton Horseshoe Tournaments came in 1998 as the result of several years of declining interest, and the pits were moved to a village park. A display of antique farm machinery moved into that spot.

Jim Rice, 4-H Agent, reported an encouraging increase in dairy cow entries for 1998. He said, "Dairy may be in trouble, but you've got to remember that it's still a $40 million-dollar industry in Delaware County. It's the county's number one business." The Fair became the first county fair in the area to offer a full-fledged llama show in that year. Also, it began to develop its own website. There were a total of 1,185 exhibitors in the commercial tent; at least 50 were turned away because of lack of space.

There were 2,613 junior exhibits at the 1998 Fair, 459 horses at the Light Horse Show and 57 draft horses; 182 sheep, 92 swine, 128 goats, 108 beef cattle, 518 entries in the art show, 831 homemaking entries; 365 flowers and 343 fruits and vegetables.

At a meeting late in the year, fair board president Don Tweedie proposed having a Fair Person of the Year award and suggested Vern Wildenstein of Chris' Flowers as the first recipient.

Wildenstein received this award, which included a plaque and much praise for his contributions and community spirit, at the 1999 annual dinner meeting. Over a period of years, the trees he planted and the gardens he created at the Fair had been much admired.

A second award, at the dinner meeting in 2000, went to Max Dutcher, "a life-long friend of the Fair" and former board member, who, in recent years, had donated his services and expertise in construction to various projects including the Fair Store and Museum.

An oversubscription in the dairy cattle barn was a problem prior to the 1999 Fair. Jean Bull, dairy cattle superintendent, reported 386 requests including 11 oxen. A few other problems were less pleasant: people swimming across the river to get in free, increasing traffic congestion at night, a few counterfeited tickets... But the 4-H Wild and Woolly Sheep Club, under the direction of Greg Graupman and Kate Henderson, was planting a flower garden, and it asked for, and got, three flagpoles for the American Flag and flags representing New York State and the 4-H.

At the annual dinner meeting in 2000, Don Tweedie retired, having served since the death of George Rich in May, 1990. His colleagues paid tribute to Don and his wife, Jean, for their leadership and hard work during a significant decade.

John Jackson, whose family had farmed in this area for generations, was elected president, and thus began the new millennium.

John's father, Graham, had been among the farmers who fought the battle to save the Fair in 1957. Graham Jackson had joined the fair board as vice president in 1958 and served in that capacity for 20 years. John's aunt, Ida Jackson Graby, had been secretary from 1959 to 1975.

Early in 2000, the board received word of a $60,000 grant from New York State towards completion of nearly $100,000 worth of work on the grandstand. To accommodate an expected increase in attendance, 1,300 seats were added, enlarging the grandstand capacity to 5,000. The roofed structure now occupied both sides of the track with dressing rooms and other facilities brought up to date.

Other construction plans included an addition to the sheriff's building and, once again, upgrading the horse show ring. More than $185,000 was spent for permanent improvements during 2000.

In May the board recorded the death of Arvetta ("Tillie") McLaughlin, who had served on the board from 1980 to 1985 and was a tireless worker in the 4-H.

As fair week 2000 came into view, it was announced that the Delaware County Fair had been chosen by the county Chamber of Commerce to receive its Arnold Banker Tourism Award in recognition of its being one of the top agriculture-based fairs in New York State. Lynne Hoppe, tourism director for the Chamber, called attention to the vast number of people attracted to Delaware County each year because of the Fair. (The attendance in 2000 was well over 75,000.) In addition to present and former local residents, the Fair beckons a growing number of second-home owners and casual visitors. Many plan their vacations around the Fair, and an increasing number of exhibitors are coming from other areas of the state.

Ray Pucci, executive director of the Chamber, remarked that the fair board's decision to beckon "big-name" entertainers promises an even bigger economic impact. Scheduled for 2001: A country music show starring Mark Chestnutt on Saturday night (rather than a mid-week performance) and "Cooter" from "The Dukes of Hazard" show midweek.

John Jackson, fair board president, declared that the annual country music show is one of the main attractions.

Statistics from the Fair of 2000 indicated that there were more applications for concessions than could be accommodated, that entries in dairy, beef, goat and other livestock departments had either risen or remained constant, that the horse show would appreciate at least another day for all of its entries, and that the art show had had a banner year.

Two new features were a bicycle give-away sponsored by McDonald's restaurants, and a Green Thumb contest.

The give-away was sparked by Gary Tweedie, who many years before had been instrumental in having bikes given away. At a board meeting the previous September he had brought the news that the McDonald clown, Ronald, was giving away bikes at other fairs, and he and Danny Hodges went forward with the project, held on Children's Day.

Tweedie, who was first elected to the fair board in 1978 and served several terms as vice president, was elected President of the New York State Association of County Fairs for 2001.

The Green Thumb contest involved teams of volunteers who planted flower beds under the direction of Robert Parsons.

A popular new attraction was the arrival of a life-size plastic cow that could be milked.

As plans for the 2001 Fair were being made, Linda Smith, Marge Warner and Rob Conklin went to the board with a proposal to move the Open Class Goat Show to the Sunday before opening day. The show would be sanctioned by the American Dairy Goat Association, the group which had sanctioned goat shows in the 1950s and 1980s.

This request and a couple of other developments resulted in a proposal at the 2001 annual meeting to expand the Fair by two days by adding the previous Saturday and Sunday. According to Fair President Jackson, a few conflicts needed to be worked out since some commercial and food concessionaires, as well as exhibitors, were involved at other fairs on those days. But, as the year wore on, the plan to expand the Fair continued to be expressed.

The annual meeting on January 17, 2001 attracted 340 persons, the largest attendance ever recorded and necessitated use of the high school gymnasium. Mainly this reflected interest in the board election at which there were four candidates for three seats.

The previous year, Gary Tweedie, who had begun working at the Fair as a volunteer in 1966, had lost his seat by a narrow margin of 53 to 51 to Pat Covell, a concession proprietor and insurance agent. Tweedie was concessions manager when the fair board, several years previously, decided to increase revenue by charging food concessionaires a 10 percent fee on their net profits. The decision was unpopular with the concession owners, but it allowed the price of Fair tickets to be kept at a reasonable level. Tweedie had said he would make an effort to regain his place on the board.

Candidates with him were John Jackson and Jon Seeley of Masonville, both running for re-election, and Kevin O'Brien of Walton, who had been a night security guard at the Fair for some 10 years. The election tally was Jackson, 202; Seeley, 183; Tweedie, 168 and O'Brien, 104.

After the vote, board members elected officers. They reinstated Jackson as president and named Tweedie vice president, the position he had held before his previous year's defeat. Joe Hoyt was re-elected secretary and Niles Wilson, treasurer.

Subsequently, pleading the pressure of work involved in his position as president of the State Association, Gary Tweedie stepped down as vice president and Roger Dibble of Hamden, a board member since 1992, was elected in his place.

At the 2001 annual meeting, the shareholders voted to amend the bylaws, changing the Fair's IRS designation to 501c, which would allow it to accept contributions that are tax deductible.

Honored at the meeting for their 45 years of promoting the Fair were Amos Finch and Myra Youmans, proprietors of Radio Station WDLA, who were retiring.

Ray Christensen of the New York State Dept. of Agriculture and Markets, who attended the meeting, expressed his appreciation for being recognized for the 35 years of involvement he has had with the fair. He said: "We are losing farmers and that has to be turned around. The fair is more than a good time."

The 115th Fair sponsored by the Delaware Valley Agricultural Society, scheduled for August 13 - 18, 2001, is the culmination of 160 years of agricultural exhibits in Delaware County. Some of them have been fair years, and some have presented almost insurmountable difficulties. As the time for the 115th Fair drew near, John Jackson reflected. He said:

"Since I was elected to the fair board in 1986, I've seen an almost unbelievable growth from about 50,000 people during fair week to 75,000, and I've seen the Fair's agricultural roots stay strong; in some departments get stronger.

"In the new millennium, I would hope that we would keep our basis of strong agricultural strength, yet expand the Fair in both number of days and in dollars spent on national entertainment acts because that will help us draw from a larger spectrum of people. I believe that spectrum will continue to broaden because as agriculture declines, people become more curious about what is a cow, what is a sheep, what is a goat....

"Even after my retirement from the fair board–since I have been active in agriculture nearly my entire life–I would hope to attend the Fair every day, every year, still learning, still enjoying. I have a strong sense of the pride shown in the Delaware County Fair, past and present, and I believe that pride will enable this Fair to continue its prosperous run for another 160 years. Furthermore, I hope that the people on the board twenty or fifty years from now have the same sense of pride that board members and Fair volunteers have had in the past and continue to have in the present."

Photo by The Delaware County Times

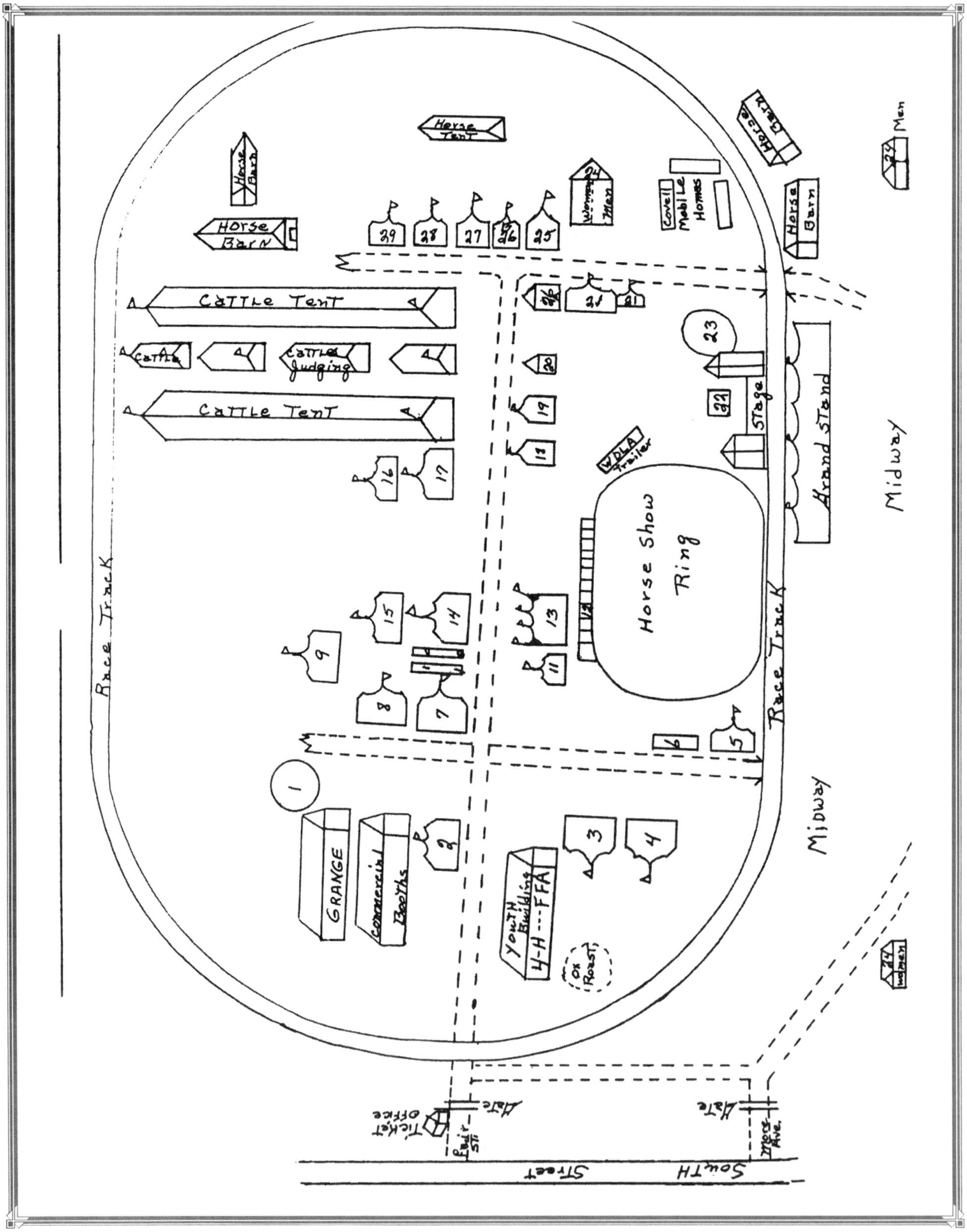

Plan of the fairground in 1963. The horse show ring is still opposite the grandstand and today's large commercial tents do not yet exist. There is a lot of open space…

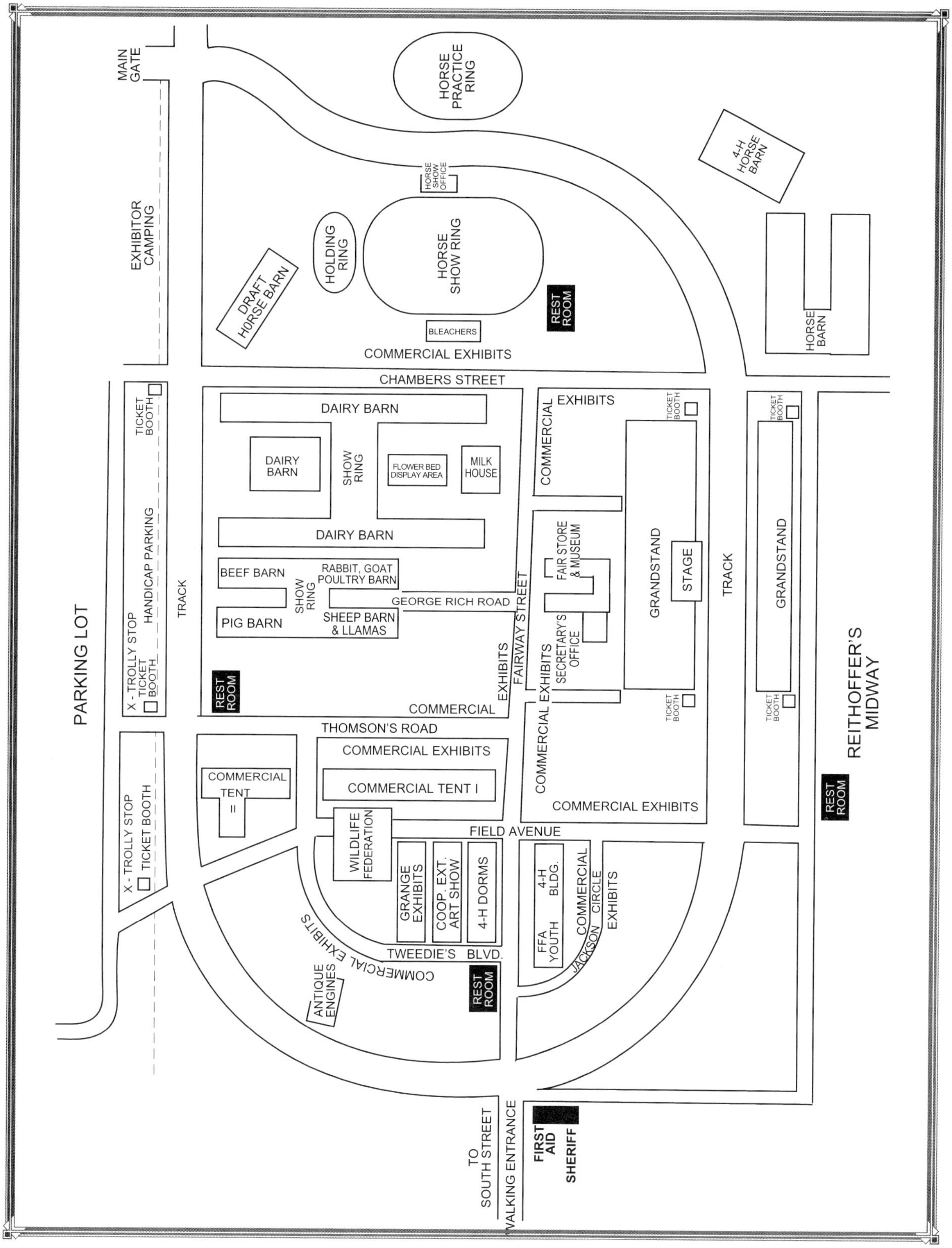

The fairground in 2001. The grandstand is now on both sides of the track and the horse show ring area has been moved and enlarged. The large dairy and cattle barns are in place and many new commercial buildings have appeared. Even the contour of the track has been changed.

Photo by The Walton Reporter

Don Tweedie of Walton, (seen here with his wife, Jean) who had been vice president, succeeded George Rich as president after his death in May, 1990. Tweedie was re-elected the following January and served for 10 years.

Photo by The Delaware County Times

One of the many projects during the 90s was a new Secretary's Office.

Photo by The Delaware County Times

The Delaware County Dairy Princess always has an important role at the Fair. In 1994, she was Jamie Burns of Treadwell, shown presenting the Premier Exhibitor Award to Janet Smith of Franklin and her grandchildren.

Photo by The Delaware County Times

Midway through January, 1996, the fairground was inundated by one of the worst floods in history. The devastation was terrific.

Photo by Robert N. Lennek

The earth was churned out all around the beautiful new grandstand.

Photo by Robert N. Lennek

Around the horse barn was another place severely damaged.

Photo by Robert N. Lennek

The floor in the cattle barn was a wreck.

Photo by Delaware Register and Revue, courtesy of Walton Historical Society

Energetic board members and volunteers, and huge cranes like this one, worked to restore the grounds in time for the 1996 Fair—and they did!

Photo by The Delaware County Times

The Fair in 1998 became the first in the area to offer a full-fledged llama show. The 4-H program had begun in 1992.

Photo by The Delaware County Times

As the decade progressed, space in the commercial tent became harder to obtain. In 1998 there were 1,185 exhibitors; others had to be turned away.

Photo by Robert N. Lennek

Some people think that food is the major reason to be at the Fair.

Photo by The Delaware County Times

Milk, that amazing product, nevertheless needs promotion. This "incredible dairy cow" is a plastic critter that actually can be milked. The kids had a lot of fun with it.

Photo courtesy of Patsy Moody

On Tuesday afternoon, after the 4-H show, the Dairy Princess, her alternates and ambassadors, turn the cattle judging ring into a Giant Ice Cream Sundae. Into a kiddy-pool, they pile some 30 gallons of ice cream, top it with delectable toppings, and everyone—between 500 and 750 people —crowd around for free ice cream. Among those pictured are Sybil Stanton, Walton; Meghan Moody, Hamden; Kelli Whittaker, DeLancey; Marianne Moody of Hamden, the 1999 Delaware County Dairy Princess; Chelsea Hanselman, Bloomville; Jackie Gray, New Kingston; and Michelle Pierce, Walton, the 1998 Princess.

For 37 years a Dairy Princess has been crowned each spring to represent Delaware County and promote the benefits of MILK far and wide. A New York State Princess also is chosen. In 2001 the two daughters of Randy and Lynette Inman of Bovina Center captured all of the crowns. Kendra, shown at left, was the 2000-2001 County Princess and became the New York State Dairy Princess for 2001-2002. Her sister, Karen, is reigning as the County Princess for 2001-2002.

Photo by The Delaware County Times

In 1999, Jean Bull, dairy cattle superintendent, reported an oversubscription.

Photo by The Walton Reporter

Don Tweedie resigned in 2000 after having served as president for 10 years. Here he is, at far right, at the annual meeting and luncheon, with his wife, Jean, at left, who had been in charge of advertising, and Richard LaTourette, a member of the board. In the foreground are Roger Dibble, another board member, and his wife, Delores.

Photo by Robert N. Lennek

Photo by The Delaware County Times

As the world changes,
and these youngsters
have children themselves,
will they be able
to take pictures like this
of those children?

Youth: The Future of the Fair

Photo by The Delaware County Times

As the years have passed, the 4-H has been cited many times as one of the Delaware County Fair's reasons for being, and the Future Farmers of America has given thousands of school children an opportunity to discover agriculture. Through good times, depressions, war years and changing ways of life, young people and their adult leaders have taken part in these programs. If the Fair is a going concern today, it is because young people were active and interested yesterday, and if there is to be a Fair tomorrow, it will be largely because of what the youth are doing today.

When the news broke, late in 1956, that the directors of the Delaware County Fair were thinking about selling the fairground to the City of New York in order to build a sewage disposal plant, many of those who immediately rushed to its rescue were 4-H leaders.

Harold Carley, George Rich, Milton Thomson, Raymond Dann, Steven Vamosy, Graham Jackson, Vanley Blackman, Michael Johnson, Frank Lamport, Ralph Gould, Robert McNee and Isabell Francisco were just a few of those associated with the 4-H in one way or another. It was a large group and they worked hard to save the Fair from the clutches of New York City.

George Rich, a Franklin farmer and breeder of prize-winning Jerseys, was chairman of the Extension Service Association at that time. He chaired the Advisory Committee which negotiated with the fair board, and later he served for 33 years as president of the Fair. Messrs. Thomson, Dann, Vamosy, Jackson, McNee and Blackman all went on to become fair board members and contributed their time and talents for many years.

Harold Carley, had been the 4-H agent for 12 years, having succeeded John Lennox in 1945. He is remembered as a short, quiet man with silvery hair, even tempered and reliable. As a 4-H official, his name appeared less often than those of the others during the tense times, but working behind the scenes, he did a great deal.

It has often been said that the farmers rescued the Fair, and that is true, but those farmers' children and families were 4-H-minded, and as the years came and went, the prosperity of the Fair depended largely upon the farmers of the future. The 4-H and a second youth group, the Future Farmers of America, have been crucial to the survival of what earlier people had considered so important to life: a love of the land, its cultivation, and an appreciation of what one-time Gov. Silas Wright in 1847 called "the first in order, the strongest in necessity, and the highest in usefulness."[1] Agriculture.

Both the 4-H and the Future Farmers of America arose during the years following World War I—the 4-H in 1918 and the FFA in 1929.

Future Farmers of America

The FFA is a nationwide, voluntary youth organization open to all students enrolled in agricultural education in secondary schools. Soon after local chapters were formed in Delaware County, youngsters began to participate at the Fair.

From the 1930s through the 1950s, nearly every school district had its own FFA Chapter. In the 1960s, however, when consolidation marked the end of rural districts and as more and more families began to leave the farm, many school districts in Delaware County did away with their agricultural programs and FFA chapters. Today there are three "traditional" agricultural and FFA programs in the county—at Walton, Delhi and Sidney, and there are two consolidated programs at the Western Delaware BOCES in Masonville and the Northern Catskill Occupational Center in Grand Gorge. In New York State, membership peaked at about 7,000 in the 1950s; today it stands at approximately 5,000, which isn't bad, considering the state of society and the various forms of competition for youngsters' time.

Children from farm families originally made up the FFA membership, but today it is open to any student interested in agriculture. The direction of interest has changed as agriculture has changed. Renwick Russell, who has piloted the FFA program in the Walton Schools for many years, points out that agriculture today is so efficient that fewer of the old-time, hands-on farmers are needed, but more and more well-educated people are crucial to the production of food. There are wide-open opportunities in marketing, processing and research, as the foods services program at Cornell University testifies.

In the school program, those who live on farms raise produce and livestock; town dwellers work at least part time on some agricultural activity. FFA offers programs throughout the year.

FFA members from all over Delaware County have been active at the Fair for nearly 75 years exhibiting home-grown cattle, livestock and horses. FFA chapters also occupy space in the Youth Building where members exhibit projects constructed during the previous year. Chapters also have booth exhibits on display showing activities held during the year and skills that students have learned in agricultural classes.

During fair week, FFA members compete in various contests, including tractor driving, dairy cattle judging and dairy cattle showmanship. There is a dairy products contest, a tree identification event, forage crops judging and a small-gas-engine troubleshooting contest. Winners go on to compete at the State Fair in Syracuse.

Birth of the 4-H

The seed that was to become the Delaware County 4-H was sown while the clouds of World War I were still drifting over the horizon. In the spring of 1918, Harry K. Lewis, secretary to C. O. DuBois, director of what was then the New York State School of Agriculture in Delhi, had an idea. He organized a vegetable club for youngsters. Fifty boys and girls joined this gardening project, which, like Jack's Beanstalk, was to grow into a high, wide and flourishing perennial. So effective was Mr. Lewis, that he was asked to organize and supervise similar groups in other communities.

Somewhere around this time, the idea of a unified youth program was being explored by various land-grant colleges, including Cornell.

Today's 4-H program in Delaware County traces its roots to both of these beginnings. Now, The Cornell Cooperative Extension is the umbrella group for what used to be the Farm Bureau, Home Bureau and the 4-H.

The First Agent

After Harry Lewis' successful garden experiment, in 1920, a Junior Extension Board was formed. Its 240 members included a wide cross-section of community life. In the following year, the Delaware County Board of Supervisors appropriated $2,500 for what was becoming known as Junior Club Work. This made it possible to hire a full-time agent. In 1922, Delaware County was among perhaps 10 in New York State to start a 4-H program.

The new agent, who arrived early in 1922, was John Lennox who served for 21 years and would gain the admiration and affection of many. He is credited with fostering the extraordinary growth of the 4-H, including the enlistment and nurturing of volunteer leaders. Camp Shankitunk (its name means "woody place"), was founded during his administration.

He was born in Ulster County and was graduated from the Kingston Academy and the State School of Agriculture at

Morrisville. During World War I, he helped farmers grow food for the war effort, and then he went to Otsego County to supervise 100 "farm cadets." He served with the Army and then returned to Otsego County in 1919 to head its newly-formed 4-H Club.

The year 1922 also marked the marriage of John Lennox to Florence Sprague, daughter of George W. and Myrtle Chase Sprague of Burlington, N.Y. Mrs. Lennox also did a great deal of work with young people. She taught in Otsego County schools and filled many kinds of leadership training roles.

After leaving Delaware County in 1943, Lennox became professor of Extension Administration at Cornell University, and from 1951 to 1975, he was extension coordinator in New England. For eight years before his retirement in 1951, he was an assistant state 4-H leader. In that capacity he developed camping programs for children and training programs for counselors at 4-H camps.

He loved children, and he loved trees. In the Town of Kortright, he developed a 180-acre parcel known as Lennox Forest; this land was later managed by SUNY Delhi. He also was a farmer and nurseryman. He operated a landscaping business and raised Christmas trees at Fojo Farm in Delhi and Morris. When he retired, he was to have managed Meridale Farms, but in 1953, he became executive director of a newly created group, the New York State 4-H Club Foundation.

John Lennox died on March 6, 1987, just a month short of his 91st birthday, and he left his estate to the 4-H. With the funds, Camp Shankitunk, near Delhi, was able to increase its camping period to six weeks. The funds also allowed youngsters who lived on farms to attend free.

One of the first things John Lennox did upon arriving in Delaware County was to foster the idea of 4-H exhibits at fairs. Boys and girls showed their animals and handiwork at the 1922 Great Walton Fair and at the Delaware County Fair in Delhi.

Although the earlier town fairs in Walton had invited entries from boys under sixteen and girls under fifteen, and a children's day was part of the schedule even in the 1880s, the first fairs were chiefly for adults, and, it could be said, for male adults. Farmers, animal breeders, poultry fanciers, and fruit and vegetable growers gathered to compete for premiums, talk about farming methods, successes and problems, have a good time with friends and cast a courtly glance at "the ladies" who were exhibiting frocks, flowers and handiwork in the Ladies' Building.

Society Begins to Change

By the time the 20s arrived, however, society was changing fast. There were not only flappers; there was also freedom–to challenge, to create, and to explore. The times were just right for a youth organization that pledged head, heart, hands and health to greater service to others and was designed to encourage independence and initiative in both boys and girls. The growth of the 4-H was both the result and the cause of some of the changes which marked the 20th century, and it certainly had a great deal to do with growth and change at the Fair.

In 1922, when John Lennox organized the first 4-H exhibits, there were 548 members in 85 organized clubs. The program was not yet called 4-H; it was Junior Club work, and it was carried on largely through the schools, with clubs in school classrooms. There were approximately 20 school fairs a year, as well as the exhibits at the Fair.[2]

In 1923, the clover-leaf symbol began to appear, and Junior Club exhibits were doing well. Gardening, poultry and homemaking were the project areas with the largest number of participants. Interestingly enough, in the 20s, there were only half as many dairy cow exhibitors as there were youngsters interested in poultry.

In 1927, the Junior Extension Board became the Delaware County Vocational Education and Extension Board, at the behest of the County Board of Supervisors, and in 1931, Youth Work became the 4-H Club Department of the Farm and Home Bureau Association. Seven members represented the two bureaus, the Board of Supervisors and the superintendents of schools.

Cooperative Extension

The 4-H was put on an even footing with the bureaus in 1937 when the joint name became The Delaware County Farm and Home Bureau and 4-H Club Association. In 1954, the names "Farm Bureau" and "Home Bureau" were dropped. These divisions became the agricultural department and home demonstration department of Cooperative Extension. The 4-H retained its own niche in this universe, and today it is described as "the youth education component of the Cornell Cooperative Extension." It represents a partnership between the Cornell Colleges of Agriculture and Life Sciences and Human Ecology, and the county Cornell Cooperative Extension Association.

By 1936, there were 1,645 4-H exhibits in Walton at what, by that time, had become the Delaware County Fair. These included a first-time entry of flowers, together with produce, poultry entries and a horse program offering $1,300 in premiums.

The first colt show was held in 1937 with Jack Briggs as the leader and 10 participants. In that year, statistics showed Delaware County as having 4,130 farms, and it reported the largest 4-H enrollment in the state—1,283 members in 100 clubs. It kept that top spot for a number of years.

There were 225 4-H dairy animals at the Fair in 1939, together with 124 chickens and other fowl, 820 garden projects, 156 clothing projects, and 17 colts. Twenty clubs marched in the parade. A 4-H band was formed in 1940.

World War II

With the outbreak of World War II, raising vegetables again became a major effort. In 1942, wartime travel restrictions made it necessary to cancel the Fair; instead, the 4-H held Community Victory Harvest Shows where bushels of garden produce were exhibited. Approximately 4,267 Victory Gardens were planted in Delaware County. Many older 4-H boys and girls went off to join the Armed Services. Shortly after Pearl Harbor, there were 697 former 4-H members serving, and in 1945, when Gov. Thomas E. Dewey visited the Fair, 37 gold stars had been placed on the 4-H flag to indicate those who had died.

With travel restrictions less tight in 1943, the fair board at the last minute decided to have its fair during the week of September 2.[3]

John Lennox Leaves

That same month, John Lennox left for a new post with the state 4-H office at Cornell University. After 25 years of service in New York State, 21 of them in Delaware County, he had become the dean. He was succeeded, in 1945, by Harold Carley, who served for 14 years.

Fair participation, generally, was high in the 40s and lower in the early 50s, a situation which should have been a harbinger of trouble, but one very likely not noticed until after the fact.

During the 40s, the numbers of 4-H exhibitors showing beef and sheep rose dramatically. Rabbits became a 4-H project in 1944, the same year that the 400 Dairy Club was formed, and goats appeared about 1948.

Halfway through the 50s, there were 192 dairy exhibits, but only 3 sheep, 26 rabbits, 29 club exhibits, and 12 entries in the parade; no pigs and no horses.

After the campaign to save the Fair had been won, the 1959 Fair "was the biggest ever for 4-H," according to the 75th anniversary book entitled *Years of Fun and Friendship*, published in 1993 as an anniversary project. The horses and pigs had returned; new exhibit records were set in many categories.

Carley and McMillen

During 1959, Cooperative Extension Agent Harold Carley left to follow in John Lennox's footsteps at the state 4-H office at Cornell, and the Leaders' Association gave him a "This Is Your Life" going-away party. He was succeeded in the same year by Ward MacMillen, who remained until 1970 when he left to teach at what was then the State Agricultural College in Cobleskill. His 11 years of service also were marked by significant accomplishments. In more recent years, Mr. MacMillen has operated the Jamaica Dream Farm in Otsego County, where he has been the proprietor of greenhouses and raised produce. After Ward MacMillen left, Brian Mateer held the position for four years, until the arrival of Jim Rice in 1974.

Hello Jim Rice

Having served for 27 years, Jim Rice now holds the record for years of service, and it has been during a time which produced some of the greatest changes in rural history. His father was a dairy farmer in

Pennsylvania, who, at one point, was a 4-H agent in New Jersey. Jim was educated at Trenton Junior College and Murray State University, Kentucky. He married a 4-H agent, the former Robbie Jean Smith and began his career with the 4-H in Tompkins County. Subsequently he became the agent in Putnam County, where he started a Youth Fair, After three years, he came here.

During the tenure of Ward MacMillen and Brian Mateer, the 4-H and the Fair experienced a period of considerable growth. The 4-H moved into its new $13,000 Youth Building on the fairground in 1963. There were 58 clubs consisting of 1,008 members in that year and they entered 3,000 exhibits at the Fair.

A new 4-H dormitory was dedicated in 1972 and named the Raymond Dann Youth Dormitory. Mr. Dann, a member of the group involved in the crisis of 1957, had become chairman of the 4-H Executive Committee in 1967. After his death in 1976, his family established a scholarship in his memory.

4-H Continues to Grow

After Jim Rice's arrival, the expansion continued. The 4-H horse barn was completed in 1976, and in the same year the Cornell Cooperative Extension moved into its present headquarters on Route 10, Hamden. The 4-H, Agriculture and Home Economics were now under one roof. For some years the 4-H had been working out of the Post Office building in Delhi and the agricultural and home people were housed in Walton, in the Post Office.

There was a county-wide meeting to organize a dairy goat club in 1978, and the 4-H horse announcing and the 5-star horse program began. Sheep and swine were added to the annual beef auction at the Fair in 1979.

Rabbits and Llamas

Early in the 1980s, a new county-wide rabbit club was formed, and guinea pigs were added to the exhibit schedule. Jim Rice received a Distinguished Service Award from the National 4-H Agents' Association in 1987. In 1988, a younger group of club members was welcomed: the Clover Buds. They are youngsters, five, six and seven years of age, who participate in special classes in areas such as the goat, sheep, swine, poultry, rabbit and horse sections. Their numbers have increased three-fold since their programs began.

Llamas became part of the program in 1992, and in 1993 the 4-H celebrated its 75th birthday.

The 20-year period between 1980 and 2000 has been one in which farming in Delaware County has diminished, and overall membership in the 4-H is considerably below its 1937 record high. The number of 4-H exhibitors at the Fair reached a low of 257 in 1993 compared to 456 in 1980. But an upward spiral then began, which has continued. In 2000, the number of exhibitors was back to 457, and there were 141 dairy cows, 63 beef, 64 sheep, 94 rabbits, 44 goats, 74 swine, 46 horses, and 89 chickens and other poultry at the Fair.

A total of 294 exhibits were entered in the Home Department; there were 166 vegetables, 61 flowers, 100 food entries, 83 items of clothing, 29 photos and 41 Clover Buds participated. Twenty members entered club exhibits, 23 marched in the 4-H Parade, there were 107 public presentations, and 91 woodworking exhibits.

Some major exhibit areas grew between 1980 and 2000, but dairy cows, vegetables, flowers, food and clothing did not. What all of this means for the future, only the future will tell.

Photo by Robert N. Lennek

First he was a school administrator, then he raised Belgian draft horses, but he was always part of the 4-H. Wendell Spear, who led the Harness and Hitches Club, drove his team in many a 4-H parade during the 1980s.

Photo by The Delaware County Times

Photo by Robert N. Lennek

"My heart, my head, and my hands..." These hands belong to Amanda Maxwell, Tiffany Hall, Travis Bennett, Kathryn Anderson, Christina Hall, Jason Little, Allison Maxwell and "Alex W.", all of Delhi.

Photo by Robert N. Lennek

Time for a little refreshment at the Fair and a raucous game of chess.

Cornell Cooperative Extension Photo

And after it's all over......

Photo courtesy of Renwick Russell

Brian Walley and Charles Fairbairn bend over a small gas engine during the Future Farmers of America troubleshooting contest. The FFA program in county schools has a long history.

Photo courtesy of Renwick Russell

An FFA tractor driving skills contest in the early 1950s.

Photo courtesy of Renwick Russell

In the Youth Building at the Delaware County Fair, a group of FFA members talk it up while waiting to see who is going to be the winner of this 1990s contest.

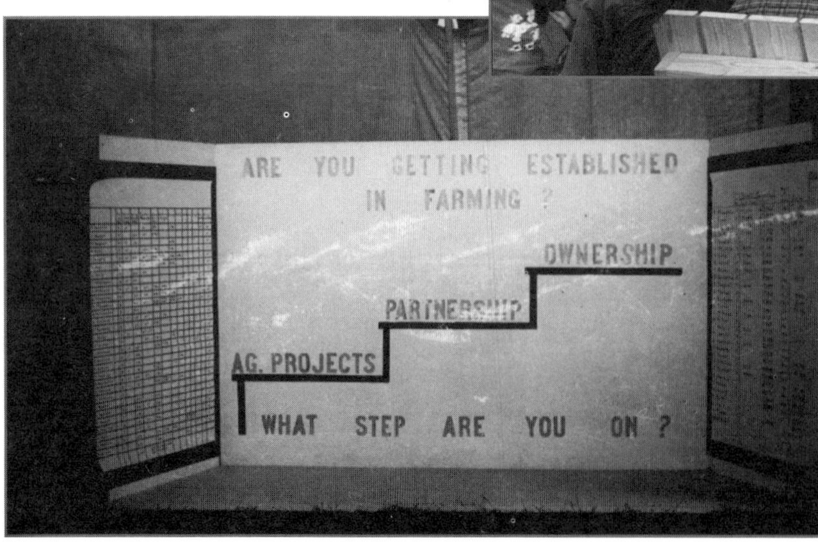

FFA bulletin board projects such as this can be seen each year in the Youth Building. *Are you getting established in farming? What step are you on? Projects? Partnership? Ownership?*

From the Cornell Cooperative Extension Archives

From the Cornell Cooperative Extension Archives

Right after he arrived, in 1922, as full-time agent, John Lennox took the 4-H to the Fair. The first exhibits were held that year at both the Delaware County Fair in Delhi and at what was then a town fair in Walton. Here are two views of the original exhibit tent at the Delhi Fair. Outside the tent, which is decorated with flags and corn stalks, stand Junior Club members with their calves.

From the Cornell Cooperative Extension Archives

From the Cornell Cooperative Extension Archives

The very first 4-H tent at the Fair in Walton. The year was 1922. On the bulletin board is a call for "men and women wanted to act as local leader in your own community." There will be a "poultry culling contest on Wednesday" and a club members' parade on Thursday. Another sign displays the well-known clover leaf: "Have you put this in your window?" The symbol was adopted officially the following year.

John Lennox at his desk about the time he arrived early in 1922.

From the Cornell Cooperative Extension Archives

John and Florence Lennox as they appeared in later years.

From the Cornell Cooperative Extension Archives

Boys who showed their calves in 1938 included Douglas Thomson, Robert Munn and Ernest Rich.

From the Cornell Cooperative Extension Archives

Until the Franklin Light Brigade was formed early in the 1950s, 4-H members interested in horses concentrated on raising draft horses. Occasionally they showed a few in the open class in Walton. Here are a couple being readied for the 1938 Fair at the farm of Lloyd Fraser in Fraser.

From the Cornell Cooperative Extension Archives

David Orton of Sidney had the Champion Holstein.

1939 Dairy Champions

From the Cornell Cooperative Extension Archives

Douglas Thomson of Walton with the Champion Jersey.

From the Cornell Cooperative Extension Archives

Brown Swiss: Mary Rose, South Kortright.

From the Cornell Cooperative Extension Archives

The first Baby Beef Club was formed in 1941.

From the Cornell Cooperative Extension Archives

An early exhibitor in the swine department was Marion Jack of Delhi, who showed this pig in 1941.

From the Cornell Cooperative Extension Archives

Another 1941 exhibitor was Richard Gifford of Masonville, who won this trophy in the draft horse show.

From the Cornell Cooperative Extension Archives

When John Lennox left for a new post with the state 4-H office at Cornell in 1945, Harold Carley took his place. Here he is in 1956 presenting a citation to Mrs. Howard Sanford for 25 years as a leader. He left in 1959 for a position at Cornell.

From the Cornell Cooperative Extension Archives

Harold Carley was highly instrumental in organizing the protest movement in 1957 when the fairground in Walton was threatened with a take-over by New York City for a sewage disposal plant. Here he addresses celebrants at the 4-H 75th birthday celebration in 1975. Standing behind him is Jim Rice.

From the Cornell Cooperative Extension Archives

At the 75th birthday celebration, Harold Carley chats with Robert Tyler of Terry Clove, Hamden, whose interest for years was Brown Swiss cows, and whose oxen are a well-known sight at the Fair.

From the Cornell Cooperative Extension Archives

David Carley and Sherrie Nesbitt with baby chicks in 1951.

From the Cornell Cooperative Extension Archives

Constance Dorland of Jefferson was one of the first dairy goat 4-H club members in Delaware County. Here she is in 1952 with her Saanen goat on her family's farm in Harpersfield.

From the Cornell Cooperative Extension Archives

John Johnson of Meridale at the 1955 Fair doing the milking machine routine in the annual competition.

From the earliest times, the 4-H parade has been a highlight of the Fair. In 1957, the Franklin Little Clovers 4-H Club mounted themselves and their calf on this vehicle and enjoyed the applause of the crowd. The Franklin Light Brigade is right behind them.

From the Cornell Cooperative Extension Archives

From the Cornell Cooperative Extension Archives

1955 Guernsey winners: Harriet Dent of Davenport and Doug Thomson of Walton.

From the Cornell Cooperative Extension Archives

Katherine St. George, the long-time Representative in Congress for this area, established junior leadership prizes for 4-H members during the 1950s. Dr. E. Ogden Bush is shown presenting the 1956 prizes to Jane Higley, Guerdon Walley and Nancy Nesbitt.

From the Cornell Cooperative Extension Archives

Ward McMillen became the 4-H agent when Harold Carley left in 1959. He is shown here with Sharon Sutherland, Sandy Bull, Randy Inman and Roger Van Valkenburg.

From the Cornell Cooperative Extension Archives

Jim Rice, who arrived as the 4-H agent in 1974, now holds the record for years of service. The 27-year period has seen great changes in rural living; at the same time the 4-H has developed many new programs.

From the Cornell Cooperative Extension Archives

Lindy Miller of Walton was blessed with a big surprise in the spring of 1964. Her Dorset ewe delivered triplet lambs. Lindy Miller Taggart is the daughter of Paul Miller who farmed for many years in Mundale. He is a descendant of Andrew Easson, one of the earliest exhibitors of sheep at the Fair. Easson is said to have walked his flock down East Brook Road from Mundale to the Fair in the early 1900s.

From the Cornell Cooperative Extension Archives

The LaTourette family has been involved in many aspects of the Fair for at least three decades. Richard and Nancy LaTourette organized the first 4-H swine club in Delaware County. Here is their daughter, Nancy, and their son, Brian, in 1985.

Top showmanship awards at the 1982 Fair were won by Dianne Sutherland and Shari Johnson. Arveta "Tille" McLaughlin, an active 4-H volunteer and a fair board member from 1980 to 1985, is representing the Eastern Milk Producers in presenting a halter to Dianne and a show blanket to Shari.

From the Cornell Cooperative Extension Archives

Photo by The Delaware County Times

This young man, pictured during the mid-80s with his rabbit, is participating in a 4-H department created in 1944. Jon Seeley, a current fair board member, has been program leader for a number of years.

Photo by The Delaware County Times

Matthew Graupman of Walton, with his eye on the judge, hopes that his entry gets a nod.

Photo by The Delaware County Times

As the Millennium arrived, Ed and Jeri Brayman of Mormon Hollow had been superintendents of the swine department for 21 years. Their children literally grew up at the Fair, showing swine, sheep and cattle. Eric Brayman, shown during the mid-80s, doing what he did very well, went on to become a Triple Crown winner, with championships for swine, beef and sheep.

Photo by The Delaware County Times

One of the newest 4-H ventures involves the care and exhibiting of llamas. This program began in 1992.

Photo courtesy of Irma and Leo Pierce

Floats, horses and wagons, equestrians, and club groups with banners, all march in the annual 4-H Parade. For the Jolly Farmers in 1980, cornstalks and hay bales were the order of the day.

People We Remember

If all of the people could be gathered together who helped make the Delaware County Fair what it is today, it would be an illustrious crowd. From the archives, here are one or two pictures of Fair leaders through the years and one or two words about some of them:

Walton Historical Society Photo

George Rich, far left, served as fair board president for 33 years, and Milton Thomson, far right was the treasurer for 21 years. Raymond Dann, second from left, was a fair board member and tireless worker for the 4-H. Board member Herbert Elderkin, known for years as The Voice of the Fair, is shaking hands with Lt. Gov. Malcolm Wilson at an event sometime during the 1960s.

George Rich

"There are a lot of good men, but every once in a while the Lord makes a gentleman's gentleman, and George Rich was one of them," said Walter Gladstone, the Andes Town Supervisor who served on the county board with George Rich.

When he died on May 23, 1990, Rich had been president of the Delaware Valley Agricultural Society since 1958, the year the "farmers" saved the Fair, and the time in between had been one of growth and success. There were many people responsible, but when the progress of the Fair is mentioned, the name of George Rich comes immediately to mind.

He was born in Oshkosh, Wisconsin on August 8, 1914, the son of Homer and Edith (Coxhead) Rich, and came to this area when he was four years old. He attended the Delaware Literary Institute and Franklin Central School, where he played center on the football team

From the Cornell Cooperative Extension Archives

George Rich, retiring president of the Delaware County Extension Association, greets the new president, Milton Ward, in 1958.

and was active in the Future Farmers of America. He was a graduate of Syracuse University.

Much of his life was given to public service of one kind or another. He was an active Boy Scout leader. He served as a member of the Franklin School Board, and for 19 years, beginning in 1955, he was the Franklin Town Justice. He was also on the board of the Delaware Valley Hospital. He became a town Republican committeeman during the 1950s, and in 1966 he became chairman of the Town of Franklin Republican Committee. At the time of his death, he was Supervisor of the Town of Franklin and vice chairman of the County Board of Supervisors.

He was a member and past Master of the Franklin Masonic Lodge 562 F. & A.M., and he was also active in his church.

George Rich was a full-time dairy farmer and breeder of the Rich-Dorr registered Jerseys. A member of the Purebred Dairy Cattle Association for 40 years, he was honored posthumously by the group in 1990 at the New York State Fair. He had been its treasurer for 30 years. He was listed as one of the outstanding farmers in New York State in 1974.

When the Fair crisis began in 1957, George Rich was president of the Delaware County Extension Association and became chairman of the Advisory Committee, made up largely of 4-H leaders. After a new fair board was formed, Rich became president and served for 33 years. The present 4-H agent, Jim Rice, has said that Rich "always made sure that the outcome was in the best interest of the children. He was out there 100 percent for the kids."

His wife, Dorretta (Gregg) Rich, was a native of Syracuse and a teacher in Franklin. The couple was married on December 24, 1938 and had two children, Margaret and Walter. After her husband's death, Dorretta Rich served out his term on the town and county boards. She died on April 1, 1994 at the age of 85.

Patricia Breakey, in an article for *The (Oneonta) Daily Star,* (January 27, 2000,) describes George Rich's last day as "a testament to the kind of life he lived. He collapsed and died while driving a fence post in his farm field after a long day dedicated to attending meetings and taking care of business related to a few of his many obligations." He had chaired a meeting of the county Health and Social Services Committee in the morning, followed by the Supervisors' board meeting at which, as chaplain, he gave the invocation. Then, at home, he did some work for the Delaware Valley Agricultural Society and went out to fix his fence. He was 75.

Walter Rich

Walter Rich, who spent his early years on the Rich-Dorr farm in Franklin, showed their prize Jerseys at the Delaware County Fair –but loved trains, grew up to be Chairman, President and Chief Executive Officer of the Delaware Otsego Corporation and its wholly-owned subsidiary, The New York Susquehanna and Western Railway Corporation. The NYS & W is a freight railroad serving upstate New York and the New York Metropolitan region.

He was born in Oneonta on January 9, 1956, the son of George C. and Dorretta (Gregg) Rich, attended local schools and received a bachelor's degree in political science at Syracuse University in

From the Cornell Cooperative Extension Archives

George Rich at the mike during the 1976 Fair.

1968. He entered the College of Law at Syracuse and received his Juris Doctor in 1971. He has maintained the family home in Franklin while also living in Cooperstown with his wife, Karine and their two children, Stephanie and Derik.

In 1966, he became General Manager and Director of the Delaware Otsego Railroad; in 1971 he assumed his present position. During his career, he has served in many capacities and received many honors.

Beginning in 1988, Rich led the effort to save the bankrupt Delaware & Hudson. His firm operated the D&H under emergency orders from the Interstate Commerce Commission for a year and a half until the D&H was sold to the Canadian Pacific Railroad.

In 1972, he was named to a committee to re-evaluate the Public Service Law as it applied to railroads. As the treasurer and past chairman of the Eastern General Managers' Association, he has

Photo by Linn Zandbergen

Walter Rich

Oneonta Daily Star photo, courtesy of Walter Rich

Both Walter and Margaret Rich showed their family's Jerseys. Here is Margaret in 1966 with Rich-Dorr Alex Suzanne, which had just been named 4-H Jersey Grand Champion.

worked to improve rail safety, organizing numerous industry seminars on the subject.

In 1993 he became a member of the New York State Public Transportation Safety Board. He currently serves on the Rail Passenger and Rail Freight task forces of the New York State Department of Transportation.

The Otsego Chamber of Commerce named George Rich "Man of the Year" in 1994, and in 1997 he received the Ellis Island Medal of Freedom. He served on Gov. George E. Pataki's Transition Team and on the Transportation Subcommittee in 1994. He currently serves on the Rail Passenger and Rail Freight task forces of the New York State Department of Transportation. In 1990, *Modern Railroads Magazine* cited the NYS&W as its Railroad of the Year.

Among other activities, he serves as a trustee of Glimmerglass Opera, Cooperstown; a trustee of WSKG Public Broadcasting, Binghamton; chairman and Eastern Region vice president of the American Short Line Railroad Association, Washington, D.C.; trustee of the J. W. Barriger II National Railroad Library, St. Louis, and a member of the Franklin Industrial Advisory Board, Syracuse University School of Management. He was elected a director of Energy East in 1998.

Milton Thomson

In 1994, when the Delaware County Fair board dedicated its premium book to Veda and Milt Thomson, they wrote that the couple had "worked long and enthusiastically for the success of the fair." True enough, but what a good example of the inadequacy of language!

Mr. Thomson, who died in 1997 at the age of 84, had been part of the Fair for a good part of his life. He was one of the leaders in the successful struggle to save the fairground when its sale was about to be finalized, and he served as treasurer for the next 35 years. He was an active 4-H participant.

Milt Thomson was born in Walton on March 10, 1913 to Charles and Mary Thomson. He was graduated from the Walton Central School where he starred on the basketball team as a six-foot forward. He married Veda Louise Alexander in 1934; they had one son, Douglas.

At their West Brook farm, which Mr. Thomson inherited from his father, the family had a herd of some 45 registered Guernsey cows and 2,000 Leghorn hens. They also operated a farm in the Town of Tompkins where they raised cauliflower, for some years a mainstay crop in this area. He was also active in the work of the Delaware County Soil and Water Conservation District, which in 1996 honored him for 40 years of service. He was also honored by the Walton Grange.

He had been president, both of the Board of Education in Walton and the Delaware County Farm Bureau, and he served on the Town Board. The Republican Party in 1989 voted him "Republican of the Year." He was also president of the Walton Co-operative Fire Insurance Company and a member of the Walton United Presbyterian Church.

Raymond Dann

After Raymond M. Dann died on November 2, 1976, at the age of 60, the fair board wrote: "His smiling face, his great sense of humor, and his skills in solving maintenance problems on the fairground will be greatly missed by all, especially those who worked closely with him in staging the County Fair each year." He had been elected to the board in 1970.

He was born in Downsville on August 27, 1916, the son of Ernest and Olive (MacDonald) Dann. He was a graduate of Delaware Academy and attended what was then the Delhi Agricultural and Technical Institute. On September 23, 1938, he married Janet MacGregor, and for 25 years they operated a dairy farm in DeLancey. By trade, he was a stationary engineer, and he worked at the Delhi College beginning in 1961. His chief interest, however, was the 4-H, and he is considered a prime mover in acquiring the land and building the facilities at Camp Shankitunk in Delhi in 1968. He was highly involved in the fairground crisis of 1957; the Raymond Dann 4-H dormitory at the fairground is named for him. He was a member of the Guernsey Breeders Assoc-

From the Cornell Cooperative Extension Archives

Raymond Dann, center back, with Mildred Johnson and Wilbur Oles, Jr. In front, Janith Kilpatrick and George Rich. All members of the 1968 4-H Program Committee.

iation, the Dairymen's League, a director of the Delhi Fire Insurance Company, Past Master of the Hamden Grange, an Elder in the Second Presbyterian Church and a member of the Delhi United Ministry.

C. O. Field

The 1920s and early 30s were years when leering side show barkers beckoned, and gambling games were almost what a fairgoer expected. Delhi, which was the County Fair until 1926, had well-publicized problems with this scene (see chapter *Walton Becomes the County Fair*). Walton, also, was not immune.

By 1934, Walton was looking for someone with "a good, clean Midway" and the fair board sent out a scout to see what could be had.

The scout was C. O. Field, who in the course of his life served in France in the First World War and in the National Guard for many years; he was Delaware County Treasurer, an engineer on the East Sidney Dam, a physical education teacher in Hobart, and principal of Walton High School. He also directed the recreation program in Walton for 20 years.

What C. O. Field found was Reithoffer's Rides, a fixture at the Fair since that time. Field was elected to the board in 1936, served as secretary until Paul G. Williams took over in 1939, and remained a member of the board until the reorganization in 1958.

His devotion to the Fair was intense, reports his daughter, Pam Field Graupman, who as a youngster watched her father hand out the trophies at the Horse Show. At the Fair, he would prowl the Midway. "If you're cheating, you're out of here," he would growl at a dubious booth vendor. They knew he meant it, and a reputation for that "good, clean Midway" became one of the Fair's assets.

Photo courtesy of Pam Graupman

C. O. Field, who brought Reithoffer's Rides to the Fair, was much in evidence at the festivities when Reithoffer celebrated its 50th anniversary at the Fair in 1984. Beside him is his grandson, Josh Graupman, who was a year and a half old at the time.

Herbert Elderkin

The 1978 Delaware County Fair premium book was dedicated to Herbert Elderkin. He was one of those instrumental in seeing that the Fair continued in 1957, and he was known for years as The Voice of the Fair because of his cheerful announcements over the public address system. He served on the fair board from 1958 to 1979.

Herb Elderkin was born September 25, 1901 in Pines Brook to Elmer and Fannie DuMond Elderkin. He was married to Iris Constable in 1927 and the couple had five children, a son, John, of Walton and four daughters Pauline E. Clair, Barbara E. Shields, Phyllis E. Howe and Jacqueline E. Baird.

Most of his life was spent on his farm at the top of Murphy Hill where he raised Holsteins. For some years he was a director of the Cornell Artificial Insemination Cooperative and a director of the Holstein-Friesian Association. Also, he was an assessor for 27 years; in 1985 Democrats and Republicans joined forces to give him a bipartisan dinner to mark his years of service. He died on March 18, 1992 at the age of 90.

Graham Jackson

After graduating from Cornell, at the age of nineteen, with a four-year degree in agriculture, Graham Jackson returned to the western Catskills to follow in his father's and grandfather's footsteps. The family, sturdy Scottish stock, had come here with early settlers, and Graham's father, William Thomas Jackson, had settled in the valley which later became the Cannonsville Dam. The family was one of those washed out of that area.

Graham Jackson first bought the farm of Morris Tripp in Beers Brook where he raised registered Holsteins and later settled on land near his father's place. After World War II, when the reservoir construction was being planned, Jackson moved to the Back River Road, adjacent to the fairground. For years he was an active Holstein breeder, for many years considered to have some of the finest stock in New York State. His herd was recognized nationally. He was a member of the Holstein-Friesian Association and the Delaware County Holstein Club.

He also had time for community service. He served for some

Photo courtesy of John Jackson

Graham Jackson and his wife, Evelyn, in the kitchen of their home on the Back River Road.

years as a member of the Walton Central School Board of Education and was a director of the Walton Cooperative Fire Insurance Company. Of prime importance, however, was his church. The family held membership in the Beerston Presbyterian Church until that area was inundated and then joined the First United Presbyterian Church in Walton. "If you can't give God an hour out of your week, you're not much of a man," was one of his frequent comments.

Graham Jackson was elected Vice President of the Delaware Valley Agricultural Society when the new board was constituted in 1958. He served in that capacity until 1978 and then for a portion of another year when he underwent triple-bypass surgery. He died on July 14, 1995, ten days short of his 79th birthday.

Ida Jackson Graby

The sister of Graham Jackson, Ida Jackson Graby was secretary of the Delaware Valley Agricultural Society from 1959 until 1975 when Uva Scott assumed that position. She is the author of a brief history of the fair association, published in 1975.

Mrs. Graby was widely regarded as a most remarkable person. She saw to it that her brother, Graham, went to college and only then went to college herself. She earned a teaching degree and then joined the Army. She was one of the individuals who broke the Japanese Naval Code (JN-25), thus helping to win the Battle of Midway.

On loan to the Navy she met Midshipman Burt Graby, who subsequently lost his arm in a Japanese aerial attack. Back home, the Graby family first operated a farm near Apex, which was bought by New York City for the Cannonsville Dam, and then settled on a farm located just past the fairground; this was later sold to her nephew, Tom.

The fair board dedicated its Premium Book of 1975 to Mrs. Graby "to symbolize the grateful

Photo courtesy of John Jackson

Ida Jackson Graby who helped break the Japanese Naval Code.

From the Cornell Cooperative Extension Archives

Kids and chickens: Ida Jackson Graby and her brother, Graham, on the family farm near Cannonsville—later washed out by the dam.

appreciation of the Board of Directors for her untiring efforts on behalf of the Fair."

Vanley Blackman

In its Premium Book of 1983, which was dedicated to Vanley Blackman and his wife, Letha, the fair board wrote: "Vanley was one of the original committee organized to save the Delaware County Fair when there was danger that it might be discontinued. He worked long and hard to get it started and keep it growing until it reached the high level it has attained today. This took a great deal of time and effort, which he gave graciously, even though he had a farm to work and an agribusiness to conduct as well."

Vanley Blackman served on the fair board for 25 years, from 1958 to 1983, and he was its secretary in 1978 and 1979, until the arrival of Arnold Froehlich.

Vanley's forerunners, the Blackmans and MacGregors, had farmed in Arabia, at the end of East Brook Road, for at least a century. When he was about eight, Vanley joined the 4-H and tells of meeting John Lennox, the first 4-H agent, at a school fair in Franklin. He raised chickens and worked at the Walton Fair. As an adult, he set up an agricultural exhibit, displaying farm equipment and labor-saving devices. He also helped to establish the tractor pulls which have become a leading attraction.

His wife, Letha, also contributed much time and energy to the Fair. Sometime after her death, Vanley Blackman married Alice (Mrs. Howard) Merrill, who has been an energetic Fair worker as well. For some years, she was superintendent of Table Settings.

Cecil Fowlston

The Premium Book of 1980 included a tribute to Cecil Fowlston of Franklin, to whom that book was dedicated along with Graham Jackson and Herbert Elderkin, three individuals who have contributed much to establishing and maintaining the Fair for many years.

Cecil Fowlston and his wife, played a major part in establishing the early horse shows. Later he laid out the ground space for the commercial exhibits. Much of the area, with new streets which were created to improve traffic, was the work of Cecil Fowlston, affectionately dubbed "Zeke." He served on the fair board for more than a decade, from about 1968 to 1979.

He was the Mayor of Franklin for some years, was a school principal, and he was the overseer for buildings and construction at Hartwick College.

Arnold and Doris Froehlich

Typical of the couples whose contributions to the life of the Fair have helped it grow and prosper are Arnold and Doris Froehlich. Over and over, through the years, certain parents have watched their children become interested in 4-H or FFA projects and then plunged head-over-heels into Fair activities themselves, leading, teaching, encouraging and nurturing. Lois and Ed Brandow and their children, the McNee, Dann, Vamosy, Inman,

Moody, Tweedie, Pierce, Smith, Darling, Russell, Walley, LaTourette, Brayman, Graupman and Taggart families are only a few of the family groups who have helped to build the fair and who continue to make it what it is. A tribute to one of them, is, in effect, a tribute to all.

Arnold Froehlich became secretary of the Delaware Valley Agricultural Society in 1980, and he held that position until 1993. He was superintendent of the horse show from 1978 to 2000. The show in 2000 was dedicated to him for contributing "more than 20 years of hard work and dedication."

He was born in Sullivan County. A resident of Walton for more than 35 years, he operated a garage and later was a mechanic and driver at the Walton Central School bus garage. He retired in 1982.

His wife, Doris, became interested in assisting at the horse show when her daughters, Roxanne and Marcie, began showing horses, and she gave many hours to the show as well as recording the minutes for the fair board. In 1994, after her death, the board dedicated its premium book to her.

The tribute stated: "A tireless lady who unselfishly donated her time and energy to improve the quality of the show. She wore many hats through the years—she was publicity chairman of the horse show, she assisted with stabling and the endless round of jobs which a horse show involves; her final contribution as chairman of the Horse Show Trophy Fund drew acclaim from exhibitors, colleagues and spectators alike. . ."

As fair week approached Doris Froehlich was on the grounds with trowel and flowers, planting and encouraging all to "fill any empty flower pots you have hanging around." She was a leader of the Equestrian 4-H Club for several years, and she continued to help even after her own children were no longer involved. The tribute to Doris continued: "Her ready smile was always welcome and her spontaneous laughter was so contagious that her bright spirit lives on in our hearts."

These Are the Fair Board Members of Today

The directors of the Delaware County Fair for 2001: First row, left to right, Joe Hoyt of Walton, secretary; Ed Rossley, Bovina Center; Jon Seeley, Masonville; John Jackson, Walton, president; Pat Covell, Walton. Second row, Gary Tweedie, Walton; Niles Wilson, Walton, treasurer; Richard LaTourette, Sidney Center; Bob Moody, Walton; Roger Dibble, Hamden, vice president; and Danny Hodges, Walton.

John Jackson

As the current president of the Delaware Valley Agricultural Society, John Jackson is the third member of his family to hold office. His father, Graham Jackson, was vice president for 21 years, and his aunt, Ida Jackson Graby, served for 16 years as secretary.

John was born on February 2, 1956, and spent his childhood on the family's Back River Road farm. He was graduated from the Walton Central School; he married the former Karen Merwin on November 23, 1974. The couple has three sons, Todd, James and Reginald. John Jackson is a custodian at the Townsend School, and Karen is a certified nurse's aid who works at the State Veterans Home in Oxford.

Jackson was elected to the board in 1986. About the Fair he says:

The Delaware County Fair has touched the lives of, literally, millions of people over the years, and it has a reputation for being a very special fair. If there were to be another crisis like that of 1957, would there be enough farmers to come to the rescue? I wonder.

The work of Rich, Thomson, Elderkin, Dann, my father—and all the others who got us here should be remembered and cherished because our agricultural roots are very precious.

I do think the future looks okay because we have a great board, but as society continues to change, there are pitfalls. Hopefully, dedication, loyalty and integrity are not things of the past.

Gary Tweedie

Gary A. Tweedie, who was first elected a director in 1978, is serving in 2001 as president of the New York State Association of County Fairs. He was elected vice president of the Delaware Valley Agricultural Society in 2001, but resigned to give full time to the state office. He was succeeded by Roger Dibble.

Gary has been an auctioneer since he was 15 years old and is still in the business. He was born in Walton, a son of the late Walter and Alice Barnes Tweedie. He spent his childhood on the farm and has lived in Walton most of his life. He is a graduate of Walton Central School and for 32 years has been married to the former Joan Slauson. The couple has seven children, a girl and six boys, and eight grandchildren.

Beginning in 1966, he operated his Walton farm in addition to holding auctions. In 1988, after selling the farm, he became a full-time auctioneer. He says:

I fell in love with the Fair as a child and still love it today. The Fair is important to me because kids are important to me. Kids are what the Fair is all about, from the competitions and learning experiences to the games and rides on the Midway. The most enjoyable part of the Fair for me is watching a child's face when they have just won a bicycle or milked a goat or petted a pig. I work hard for the Fair for two reasons: I love fairs and I want to see our Fair remain one of the top agricultural fairs in New York State.

Roger Dibble

Roger Dibble, who was elected vice president of the fair board in February, is the proprietor of Dibble's Garage in Hamden and is also employed by the Hamden Highway Department. He was born in Margaretville on March 23, 1955, the son of Stanley and Virginia Dibble. The family moved to Hamden in July 1955.

On August 18, 1973, Roger married Delores Jabar, the daughter of John and Mabel Jabar of Oneonta. The couple has two daughters, Lisa Jones and Stacy Dibble, both of Hamden, and a grandson, Tyler.

Roger has been associated with the Fair for nearly 25 years, having begun as a member of a group which volunteered to improve the track. He was elected to the board in 1992.

Why is the Fair important? He says:

I feel that the Fair is very beneficial to our area because it provides a place for all kinds of learning. Also, it's good entertainment for low dollars; it's good for the entire Delaware County economy, and it teaches kids a great deal that they need to know about agriculture and healthy competition.

Joe Hoyt

L. Joseph Hoyt, secretary of the fair board, was elected to that position at the annual meeting of the Delaware Valley Agricultural Society in January, 2000. He first became a board member in 1998.

A native of Mundale, Joe is the third generation of a large

family of Hoyts who farmed in that once-thriving hamlet on East Brook, north of Walton. He was born May 7, 1955 to Bruce and the late Barbara Hoyt. His uncles, Walter, Hilton and Larry were widely known for their musical abilities, and Walter operated a well-known sap house. Joe began to show dairy cattle at the Fair when he was five years old.

After the elder Hoyts ceased farming in 1972, Joe became associated with Vanley Blackman in his Jamesway barn equipment business, helped with the equipment display at the Fair and worked on the Blackman farm in Arabia. He was also active in the tractor pull events. He has been a member of the tractor pull committee since 1974.

Joe Hoyt says:

The Fair is very important to me because it is such an excellent place to educate non-farm people about the life of the farmer and show how produce and other farm products are made or grown.

It is also important because it keeps young people interested in agriculture so that maybe they will become farmers themselves. The Fair is also a good place for farmers to get together, talk about their farms, and have fun with other farm families who they don't normally see during the year.

Niles Wilson

Niles Wilson has been treasurer of the Delaware Valley Agricultural Society since the resignation of Milton Thomson in 1993.

Richard LaTourette

One of the families that has kept things spinning at the Fair for a long time is that of Richard and Nancy LaTourette of Trout Creek. Both Richard and Nancy, and their children, Brian, Nanette, Jason, Aaron and Erwin, have been active in raising and showing beef, sheep and swine for many years.

Richard was elected to the fair board in 1982 and has served in various capacities. He was vice president for several years in the mid-90s and served as secretary in 1998 and 1999.

He has been involved with the Fair since childhood. "I have been to the Fair every year since I was a child, and except for one or two years, I have been an active participant all of that time." He showed cattle as a 4-H member and has worked with the tractor pulls and tugs of war.

Why does he consider the Fair to be important?

The Fair is an educational atmosphere that promotes a learning experience for the whole family. It is youth oriented, provides a positive outlet for kids in Delaware County and beyond. In so doing, it becomes a family affair, bringing families together in a social atmosphere as they work to achieve positive goals.

Danny Hodges

When he was sixteen, Danny Hodges began to work on the farm of Graham Jackson, who was then president of the Delaware Valley Agricultural Society. Working at the Fair went along with the job, and Danny has been part of the Fair ever since. His was not a farm family, but in the course of time he showed Holsteins and pitched in to do many kinds of useful things.

Danny Hodges was born in Walton on September 30, 1960, the son of Patricia and the late Paul Hodges. He married the former Vinnie Sherman in 1981, and the couple has three children, Jillyn Reed, Amy, and Brad. He is the proprietor of the J. C. Penney Store in Walton and also is service manager at Breakey Motors. He has been a fair board member since 1998. Mr. Hodges says:

To me, the Fair is a way of life here in Walton because it contributes so substantially to the economic development of the area. It plays a large part in keeping the community alive. In addition the Fair is almost like a family reunion. You get to see friends and relatives that the busy pace of life these days doesn't let you set eyes on all year!

Ed Rossley

Edward Thomas Rossley, who was elected to the board in 2000, is the proprietor of Discount Tire Company in Delhi. He was born in San Antonio, Texas, on November 11, 1960, the son of Edward F. and the late Joan Fynaut Rossley. When his father retired from the U. S. Air Force in 1970, the family settled in Bovina Center, and Edward was graduated from Delaware Academy in 1978.

He attended SUNY Cobleskill, majoring in agricultural engineering. Then he decided to transfer to to SUNY Delhi where he graduated with a major in automotive mechanics in 1980. His wife is the former Gretchen Warner and the couple has three children, Elizabeth, Mark and Kevin. When asked about the Fair, he says:

I think the Fair is important for a number of reasons. We can learn better ways to produce, and we can watch the youth grow in confidence as they share in friendly competition in 4-H projects, showing animals, horse shows, art exhibits, and tractor pulls, just to name a few departments.

My goal for the Delaware County Fair would be to have many more people enjoy the Fair. I would challenge each person to experience the great things the Fair has to offer. It has been a privilege to participate in showing animals, selling goods, announcing tractor pulls, and serving as a director.

Jon Seeley

Jon Seeley has been a board member since 1993. He has been active at the Fair since showing cattle as a youngster and has been a 4-H leader for a number of years, his main interest being rabbits and poultry. He was instrumental in creating a "livestock" barn to house various species instead of only one.

Bob Moody

Robert Moody was elected to the board in 1999.

Pat Covell

Pat Covell was elected to the board in 2000.

Part II

Under the Tents
&
Inside the Track

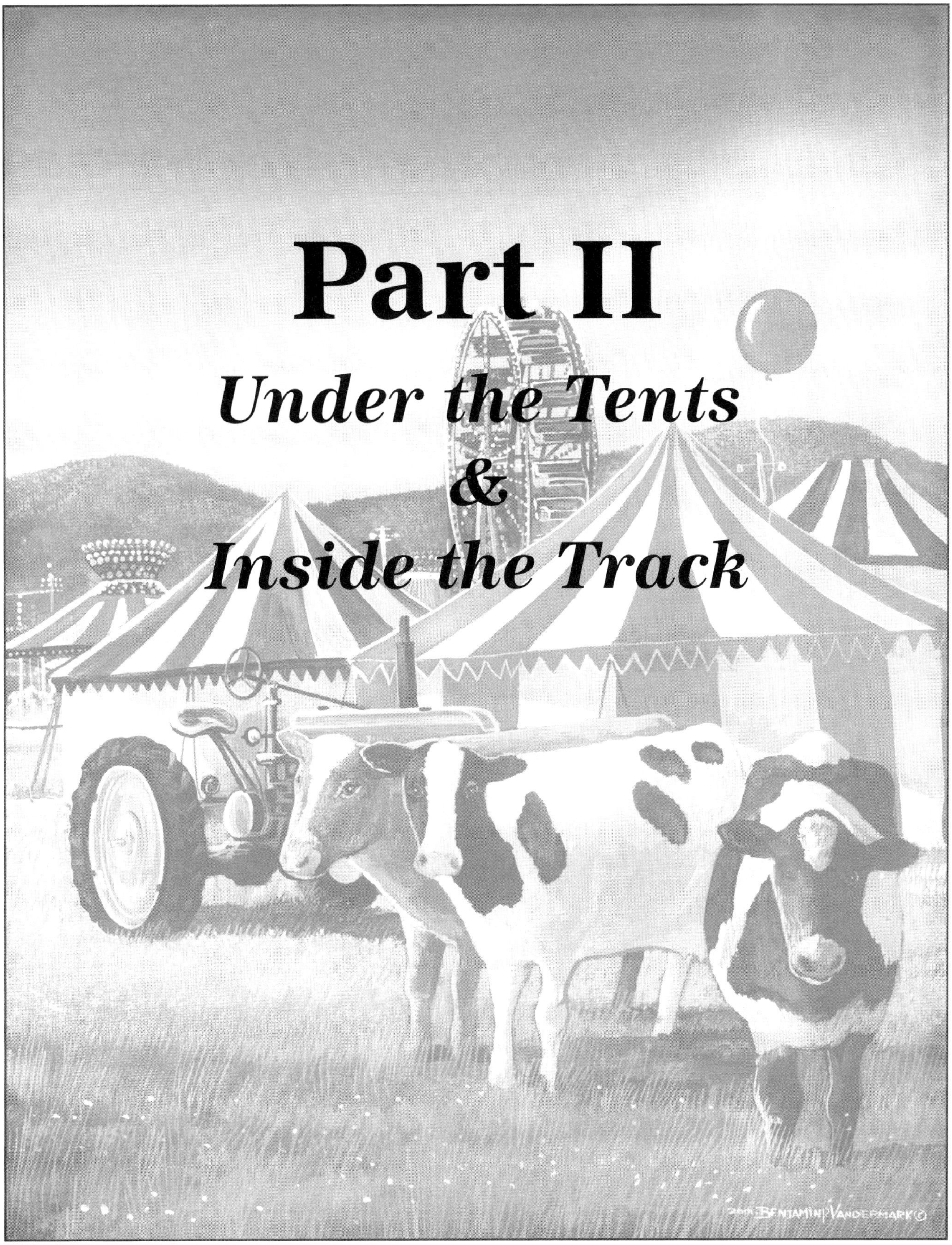

Dairy Winners and Quality Beef

Rows and rows of cows and beef cattle standing side by side on sweet-smelling hay in the stalls of the big Fair barn on a sunny summer morning.

Holsteins, Jerseys, Guernseys, Ayrshires
Durhams, Herefords, Galloways and Angus

A picture that captures the spirit of farm life and distills the hopes and expectations of more than a century. But look at these cattle. They're here today. Gone tomorrow?

When the first county fairs were held in Delhi, there wouldn't have been any Jerseys, Galloways or Angus lined up waiting to be judged because those breeds weren't imported in any great numbers until later in the century. But several important kinds of dairy cows and beef types had been around, on the family farms, and they might have been seen at the 1841 Fair.

- Durhams, both milch cows and beef animals, had come from England as early as 1783. We know them as Shorthorns. The first registered bulls were brought into New York, to the Genesee Valley, in 1817 by a Samuel M. Hopkins. From 1820 to 1850, the supply of Durhams increased rapidly, and in the 1840s they were popular. Today Shorthorns are separated into distinct dairy and beef types, each with its own herd book.[1]
- Devons (strictly speaking, "North Devons") had been

around since English colonists brought them to America about 1623. For years a Devon was "the old red cow" which early settlers depended upon for milk, beef, ox teams and leather. Devon oxen are still to be seen on local farms.[2]

- Ayrshires, from Scotland, started to arrive in 1822 followed by others between 1837 and 1858. Ayrshires and Jerseys are reported to have been the most popular breeds just after the Civil War.[3]
- Holsteins. An early version might have been seen during the 1840s. In the beginning Holsteins were known as "Dutch Cattle" and were one of the earliest milkers in this area, having been brought here by the Dutch between 1621 and 1625 when they settled the Hudson and Mohawk Valleys. But when the British came, so did an avalanche of their cows and until 1860 or so the Dutch Cattle were a very mixed breed. It was Winthrop W. Chenery of Massachusetts who was responsible for the first permanent herd of purebred Holsteins in this country. The breed came to be called "Holstein," interestingly enough, as the result of a "mistake." Chenery himself referred to his animals as Dutch Cattle, and in an article for the United States Department of Agriculture, that is what he called them. But apparently someone at the USDA changed the name. Chenery and others went along with it, however, and presently the cattle were officially "Holsteins." It could be said that the USDA named the breed. The joint name Holstein-Friesian did not begin to be used until the late 1800s, but more about that later.[4]
- Herefords came in 1817, from England, when the famous statesman, Henry Clay, took a bull and two heifers to his farm in Ashland, Kentucky. In New York State, Erastus Corning and W. H. Sotham imported 21 cows and a bull in 1839 and 1840. Herefords were exhibited at the New York State Fair in 1844.[5]
- Guernseys, which originated on the Isle of Guernsey, off the coast of England, more than 1,000 years ago, began to come into this country about 1830, with one cow and a bull as the foundation animals. The original imports were taken to Gen. Moody A. Pillsbury's farm in New Hampshire where they were billed as the Pillsbury Bull and Pillsbury Cow. The New Hampshire Guernsey herd book developed from these two. The American Guernsey Cattle Club was not organized until 1877, however. J. C. Penney was a noted Guernsey breeder and two Guernseys went to Antarctica with Admiral Byrd's expedition in 1933.[6] But Guernseys ran into a bit of trouble and today they are on the "watch" list of the American Livestock Breeds Conservancy.

Thus, at the early fairs, one might have seen Durhams, Devons, Ayrshires, Dutch Cattle and maybe, but not altogether likely, Herefords and Guernseys. There were no separate classes for dairy animals and beef cattle until much later.

The first Jersey cows were imported from the British Isles in 1850, and they immediately became popular because of their rich milk. Flora, a famous Jersey, is reported to have given enough milk to produce 511 pounds, 2 ounces of churned butter in 50 weeks, a record for her time.[7]

Galloways, small, hardy, polled black cattle, whose written records go back as far as 1530, are native to southern Scotland. They arrived in Canada in 1853 and came to the United States in 1866 where they became tremendously popular at all of the livestock shows. But as the Aberdeen Angus began to rise in favor, the Galloways declined in numbers except in the Western states.[8] They, too, are on the "watch" list.

The Aberdeen Angus, black and without horns, were comparatively late-comers. They were first imported in 1873 from Scotland, with a second lot arriving in 1879. Their reputation for producing top-quality beef made them soar in popularity, especially during the 50s when they were a strong challenge to Herefords and Shorthorns.[9]

Not long after the various breeds were imported and their numbers began to increase, a breeders' organization arose for each of them.

They include: the American Jersey Cattle Club, 1868; the Association of Breeders of Thoroughbred Holstein Cattle, 1871; the Ayrshire Breeders' Association, 1875; the American Guernsey Cattle Club and the American Galloway Association, 1877; the Brown Swiss Cattle Breeders' Association, 1880; the American Hereford Association, 1881; the American Shorthorn Association, 1882, covering both milking and beef Shorthorns; and the Aberdeen-Angus Association, 1883.

The years covered by the formation of these groups, roughly from the end of the Civil War to almost the turn of the century, was a time of enormous change. All kinds of new cows and beef cattle were coming in, new kinds of agricultural machinery was appearing and the pervading attitude was one of "onward and upward".

Right in the middle of this cultural scene, and no doubt influenced by everything that was going on about them in the way of agricultural events, were C. B. Bassett, Joseph and Robert Harby, William C. Doig, Henry W. White, William C. Rogers, E. R. Howland and Alexander Tweedie, the prime movers in the organization of the Delaware Valley Agricultural Society in Walton in 1887.

As some of the more prosperous and forward-looking farmers became aware of the new opportunities that purebred cattle offered in terms of higher profits and personal satisfaction, specifications for exhibiting began to change. In the Delaware Valley Agricultural Society's premium book of 1887 the cattle department, headed by David R. Scott, had a classification called "Thoroughbreds" (which was, of course, a misnomer since that is a breed of horse. But it was a much-used term at the time.) Under the "Thoroughbred" heading were classes for "Short-horns, Devons, Holsteins-or-Friesians and Jerseys."

Guernseys were among the Grade animals that could be shown. ("Grade: At least half pure blood or more"), together with "Short-horns, Devons, Holsteins-or-Friesians, Jerseys, Ayrshires" and "natives," which were defined as "less than half bloods."

With the appearance of the hyphenated title, Holsteins-or-Friesians, it is time to tell the story of Solomon Hoxie of nearby Otsego County who caused much turbulence at a pivotal point in the history of the Holstein breed, and at the same time is credited with exerting "an influence for the betterment of the breed second to no other man in the country."[10] Here is an important chapter in national breed history which was written right in Walton's back yard.

The years between 1877 to 1885 are usually considered the period of important development for the big black-and-white cow which has given farmers such large returns in milk production. Winthrop W. Chenery, who had imported the first animals from the north of Holland, and his followers including Gerrit S. Miller of Peterboro, New York who had brought over cows from Friesland in 1869, were the main participants in a breed association formed in 1871.

Between 1871 and 1875, Thomas E. Whiting of Concord, Massachusetts, began to import Dutch Cattle also, and a fine feud developed between Whiting and Chenery.

Enter Solomon Hoxie of West Edmeston, who, with a group of young farmers formed the Unadilla Valley Stock Breeders Association. Presently they bought animals from Whiting, and in due course they formed a breed association called the Dutch Friesian Association of America. It inherited, or at any rate continued, the feud with the Chenery group and for the next decade, differences concerning numerical representation, the name of the breed, and the source of pure breeding stock resulted in bitter, explosive and cantankerous controversy.

Hoxie, whose parents were Quakers, had a small farm and on the side was a local preacher for the Free Will Baptist Church in West Edmeston. Apparently he was becoming interested in finding better stock than the area then provided, and as manager of the livestock business, the Unadilla Valley Stock Breeders' Association, he went to Holland several times to select cattle. He was said to have had a good eye for superior type in the animals.

He served as secretary of the Dutch Friesian Association throughout its existence from 1874 to 1885. From all accounts, he was a strong-minded, articulate man with definite opinions and a persuasive way of expressing them.

His clash with the Holstein people was, first of all, based on claims that Holstein numbers had been unfairly increased and that animals had been admitted from too wide a territory.

The group also argued that "Holstein" (which is an area in Germany) was a completely incorrect name for a Dutch animal. They campaigned mightily with the slogan ***A Pure Register and a Correct Name***, and for years the wrangling and jockeying for public favor and public supremacy devoured many columns in the agricultural press and spilled over into other arenas.

As salient as these points were, however, it was a fourth item, the idea of an Advanced Registry System, that would truly benefit the breed and give Solomon Hoxie a place in history.

This plan involved a dual-record system, a separate pedigree registry and a performance or "Advanced Registry." Any purebred cow could be registered, but only cows that met or exceeded certain requirements for production would be entered in the Advanced Register. In future years all of the major dairy associations would adopt this way of recording their best producers, and in that way they would have a concrete reference for improving the breeds.

A series of circumstances, including an outbreak of hoof-and-mouth disease in Europe and falling prices at home, combined in 1885 to make a union of the feuding organizations particularly desirable. The breed became, officially, Holstein-Friesians. Theron G. Yeomans, a Wayne County, New York breeder, was the first president of the new Holstein-Friesian Association of America, and Solomon Hoxie became its Superintendent of Advanced Registration, a position he held until he retired in 1905 at the age of seventy-five.

The term Holstein-Friesian appeared in the premium book when these cattle were shown at the second Delaware Valley Agricultural Society fair held in 1888. Guernseys and Galloways were added to the "Thoroughbred" list. John S. Holley had become superintendent of the cattle department, and he served in that capacity for several years. He was succeeded by E. B. Howland who had the job for nearly a decade.

The 1900s, until well after World War I, were a time of great variety in the cattle barns. In 1903 one could still show "Grades—Jerseys, Devons, Short Horns, Holstein-Friesians, Guernseys, Ayrshires, Herefords, Galloways and Red Poles" (Red Polls). By 1909, however, the classes for grades had been eliminated, and the Holstein-Friesians were being called simply "Holsteins".

In 1916, in addition to the old favorites, there was a class for purebred Red Polls, a dual breed that had begun to be imported in 1873 and by the early 1900s was enjoying a short but strong burst of popularity. There were also Dutch Belted, Dexter-Kerry, Columbian, and what in future years was to become one of the top milking breeds, Brown Swiss. This "Big Brown Cow," one of the oldest dairy breeds, was first imported in 1869. She was originally considered a dual-purpose animal but a high potential for milk production enticed breeders, beginning in 1907, to emphasize her dairy-cow qualities."[11]

That 1916 Fair, with its long list of cattle signed up to compete for premiums, was typical of the period, but soon the scramble for supremacy was to thin the ranks to a few strong front runners. Production, marketing prowess, breeding innovations, organizational leadership—and fashion—were to cast one, then another breed, into the spotlight.

During the time when butter was an important Delaware County product, Jerseys and Guernseys were numerous at the Fair. Their milk had a high butterfat content and turned to butter easily. But the Holstein Association was working hard to improve its breed, and as early as 1887 a Holstein cow, Clothilde, confounded all the experts by winning $200 and a silver cup for being the most productive cow for both milk and butter.[12] It was the beginning of the road upwards for the "black-and-white" who, as the years passed, filled more and more of the Walton Fair dairy barn stalls.

The early 20s were hard for the region's dairy farmers and hard for the Fair which consistently was losing money. Early in 1924 fluid milk was pegged at $2.33 a hundredweight.

The stock market crash of 1929 at first was not felt in the farmland, but by 1931 cattle prices had collapsed and farmers girded themselves for a difficult time. The 30s was a period of large condemnations of tubercular cattle, a process which ruined many a small farmer, and there was a growing unrest among dairymen who were at the mercy of price-cutting milk producers' associations. Faced with the threat of an influx of cheaper milk from western sources and the creation of huge dealer combinations, Delaware County farmers held a succession of "unity" meetings, which, true to predictions, went nowhere.

With the outbreak of World War II, the demand for milk skyrocketed; farmers raced to keep up with the demand, and the stalls at the Fair were full, especially with the high-producing black-and-white Holsteins, for whom the newly enacted 3.5% milk standard was a distinct triumph. The Fair was curtailed for only one year during the War. By this time health certificates were being required, proving that animals had been vaccinated for tuberculosis, Bang's disease and shipping fever. Trophies were becoming more numerous: The Delaware County Jersey Breeders Club was awarding a President's Silver Loving Cup for the grand champion Jersey female; the Otsego-Delaware Ayrshire Club was giving a similar cup to the Delaware County breeder showing the grand champion Ayrshire female. The Better Milking contests and the dairy herd improvement program were begun.

With the end of the 40s came the disappearance of the old homestead bull as artificial insemination took hold. The fair board in 1949 gave $50, $40 and $30 in prizes for the best educational exhibit involving artificial breeding. A new job had been created, that of artificial inseminator, who went from farm to farm. Suddenly top sires were available on a widespread basis.

The 50s was a period when Holsteins reached a new peak of popularity. By the end of the decade, out of every 100 dairy cows registered in the United States, 64 were Holsteins.[13]

In 1958, John Elderkin, Fair secretary, reported that there had been "the biggest cattle display in history with 300 registered head." He added, "We had to have three extra tents for cattle this year" with 60 more animals on the grounds than any other year in the past.

During the 60s there were unprecedented changes in feeding, environment and production, and as cattle changed in appearance as the result of these practices, judging standards at the Fair changed also. During this decade Dr. George Trimberger of Cornell, who had done a great deal of work in the field of dairy cow type and body conformation, was enlisted by the Holstein-Friesian Association to develop a completely new style of cattle scoring which became known as "Descriptive Classification".[14] The system, which was widely adopted in principle, served as an incentive to consider not only production capacity but conformation as well. To a degree it spoke to the tendency of some breeds to opt too much for "fancy" animals to the detriment of production. Possibly because some breed associations took this situation to heart, there started to be an increase at the Fair of "colored" stock such as Jersey, Guernsey and Brown Swiss which had become almost dangerously scarce.

Dr. Trimberger was the cattle department judge at the 1961 Fair.

The 1960 premium book lists a number of trophies being newly awarded in the cattle department, several of which were given over a number of years:

- To the first prize cow in the type and production class, by the Margaretville Chamber of Commerce
- To the best Ayrshire female by the New York Southern Tier Ayrshire Club
- To the exhibitor of the best Dairy female of all breeds, in memory of William E. and Herbert D. Henderson
- To the best Jersey female owned and bred by the exhibitor, by Mrs. George T. Rich of Hobart

In 1962 the Delaware Valley Agricultural Society honored livestock breeder and former fair board president Peter Chambers by awarding a trophy in his name to the 4-H Champion Female of all breeds. A new trophy was awarded each year until 1964.

Ralph Taylor was department superintendent from the late 50s until 1961. He was succeeded in 1962 by Carroll Maurer of Franklin who served for 12 years.

The department had a much-expanded committee in 1964: Carroll Maurer, David Orton of Sidney, Milton Thomson of Walton, Robert Tyler of DeLancey, Mrs. Cecil Francisco of Sidney Center, Graham Jackson of Walton, Richard Gifford of Deposit and Robert McNee of Delancey.

Stall decoration prizes began to be given in that year. Decorating the stalls was an appealing project for the exhibitors, visitors, and the cattle department committee alike. After a while the stalls were decorated around a theme for the year, always centered around some aspect of farm life. Some of these have been: *Four Seasons of Farming, Farming Is Our Business, Farming Through the Years, Farming Is a Family Affair, Families Make Farming Fun,* and in 2000, *Champion Cows and Champion Kids.*

The first Open Show for Beef Cattle was held in 1965.

During the mid-60s the total premiums for the department ran around $3,900.

In 1975 the department was divided into Dairy and Beef Cattle divisions.

The dairy superintendent in 1975 and 1976 was Myron Coulter of Hamden, and Charles Hoffman of Walton was in charge of beef. There were four beef categories: Hereford, Aberdeen Angus, Shorthorn and "other".

In the 1975 premium book the Purebred Dairy Cattle Association Code of Ethics appeared for the first time and although this is no longer printed in full, the Code comprises the rules which govern all cattle exhibitors at the Delaware County Fair.

This Association is 62 years old. It was formed as the result of a cooperative dairy display at the New York World's Fair of 1939 featuring a milking merry-go-round called a Rotolactor. During the planning, the breed groups felt the need for some sort of forum for ideas and joint action. The following year, an organization meeting was held with the five major dairy breeds participating—Ayrshire, Brown Swiss, Guernsey, Jersey and Holstein. Shorthorns were added later.[15] Through the years, the PDCA has had a tremendous influence in developing standards, setting policy and formulating ethics.

Included in the list of unacceptable practices mentioned in the Code of Ethics are such things as:

- Misrepresenting an animal's age
- Administering drugs
- Changing the color of hair
- Using alcoholic beverages in the feed
- Doing various things to udder or teats to increase milk
- Inserting foreign material under the skin
- Striking the animal to cause swelling in a depressed area

You wouldn't think people would try to cheat like this, would you, but they do!

Richard Niemczyk of Walton was dairy superintendent from 1977 to 1980; during these years, August Anderson of Long Eddy, Donald R. Rimbach and Donald Walsh, both of Walton, took care of the beef.

Jean Bull of Walton took over on the dairy side in 1981, followed by Dixie MacGibbon of Walton in 1985. Sometime during the 90s, Jean Bull returned and served until 2000 with Bert More of Hamden as show chairman.

Don Tweedie of Walton began a term as beef superintendent in 1981 which lasted until he was succeeded by Jim and Lindy Taggart of Walton in the early 90s.

The 70s and 80s can only be described as extraordinary. It was a time of government subsidies, newly commercialized embryo transplant techniques, spiraling international exports, and high prices for cattle. It was a good time for farmers. It was also a good time for livestock investors who came from the ranks of bankers, Wall Street brokers and rock stars like John Lennon who had a local interest. Through pooling cattle ownership in limited partnerships, the investors were able to bid for the top genetics being developed by the breeds. During this time, and because of the favorable investment climate, some of the nation's top dairy stock went through the Walton Fair. Among the herds that showed them were *Dreamstreet Holsteins, Tyrbach Holsteins* and *Catskill Holsteins.*

Dreamstreet was started by George Morgan of West Brook, Walton, in 1975. He had been a dairy farmer in New Jersey and Walton, and had been showing cows at the Fair as an individual before forming the investment group which then exhibited many top-grade animals. In 1979, he sold *Dreamstreet* to Frank Wood, a Walton attorney, who had been interested in cattle for some years. Morgan then continued to show dairy cows under the *Tyrbach Holstein* banner. One of these was the famous home-bred cow, Tyrbach Elevation Twinkie (EX-97). (The figure represents an "Excellent" category wherein the cow rated 97 out of 100 possible points.) She was the only Holstein to be Grand Champion, in the same year, of the three United States National Shows and the Royal Winter Fair in Canada.

From Frank Wood's *Dreamstreet* Holsteins came many distinguished cows which were shown at the Fair from 1979 through 1987 when *Dreamstreet* began to wind down because of changing conditions. It was usually Premier Exhibitor of the Fair since its show herd was large and included national-caliber animals.

For example, in 1980, at the Hilltop Hanover Sale, which set new records for prices and volume of sales, *Dreamstreet Holsteins* paid the then top price of $300,000 for Pammies Citation Paula-ET, the eighth generation Excellent from a family that went back to top sellers in 1968.

Its best year was 1983. Featured were two four-year-olds, MillervaleUltimate Rosalyn and Campbell-Hollow Ultimate Kate, each EX-95. Rosalyn and Kate were eventually Grand Champion and Reserve Grand Champion at the 1983 World Dairy Expo in Madison, Wisconsin, Rosalyn being named Supreme Champion of All Breeds in North America. Also in 1983, Howard-Home Valiant Eva was undefeated as a two-year-old and unanimous All-American. She was a particular favorite of Frank Wood and his colleague, Robert Harby.

Catskill Holsteins, was owned by the late Carl Hulbert and by Bruce Pierson, both of Walton. Hulbert, who had worked with cattle all of his life, had spectacular success with Shadowcliff R A Gina (EX-97), who was shown as a two-year-old at the Delaware County Fair. After the death of their friend, Carl, Frank Wood and Robert Harby donated a permanent, revolving trophy and cabinet in Hulbert's honor. It is named for Valiant Eva and Shadowcliff Gina.

The Gina-Eva Trophy is awarded to the best two-year-old each year at the Fair.

Because there are two different classes, one for junior (younger) two-year-olds and one for senior two-year-olds, after the winner of each class has been selected, the judge compares only those two animals. The winning cow is photographed with her owner and their names are engraved on a plaque on the trophy. In the 4-H section of the Cooperative Extension building in Hamden, a special cabinet was built which houses the trophy during the year.

The list of Gina-Eva trophy winners since the trophy was instituted in 1998, is as follows:

1998: Combination Astre Judy, owned by the David Post family of Pineyvale Farms, Stamford.

1999: Whoop De Do of the *Dairysmith Holsteins* owned by the Don Smith family, Franklin.

2000: Holland Heights Star Humble of *Schenevus Creek Holsteins* owned by Jeff and Paula Darling.

An inscription on the trophy reads: "Hopefully, this trophy will inspire more dedicated showmen like Carl to develop young cows of this caliber."

There have always been top-notch breeders to show dairy cows and beef cattle at the Fair. Half a century ago they included Herbert Elderkin, Peter Chambers, Merv Conklin and Graham Jackson with their Holsteins; Milton Thomson, Guernseys; and George Rich, Jerseys.

In more recent years some of the premier breeders have included Hugh Sutherland of West Delhi, who for 15 years was Executive Manager for the New York Holstein Association and served as a judge for three decades; Marian and Bill Inman of Bovina Center, Holsteins; William Walley of Walton, Guernseys and Ayrshires; David Betz of Stamford, and Ralph and Alice Taylor of Walton, Jerseys; Lester Tyler of Milford, Brown Swiss and Jerry Merrill of Downsville, Milking Shorthorns.

During the 1980s, farming was profitable. When cattle prices were high, when embryo transplants and artificial insemination were skyrocketing the cattle industry, when overseas business was flourishing and government subsidies were easily available, the cattle barns at the Fair were overflowing. The 1958 record of 300 entries was broken repeatedly.

Like every high-flying market, however, this one, too, crashed as the playing field changed, the tax laws were revised and the law of supply and demand asserted itself as customers became competitors. As a new millennium began, however, the trend line began to go up again, according to Bert More, who has been the cattle show announcer for more than 30 years and takes over as superintendent for 2001. At the Fair in 2000, there were at least 400 animals in the barns which were built to hold 368, and there were a number of new exhibitors, he reports.

"Dairy farming is a calling," Bert More believes, and there are still young people who are being called to be farmers. Two-income families who love the land are turning up more and more, he insists, with a desire to breed and show animals not as a way to pay all the bills but to live a good life. He is optimistic about farming as more than a business proposition.

Photo by The Delaware County Times

Photo by Robert N. Lennek

The dairy barn at the Delaware County Fair.

Photo by The Delaware County Times

Happiness is a day at the Fair with a good cow.

Photo by Robert N. Lennek

When I grow up, I'm going to have a whole herd of cows, just like this!

Photo by Robert N. Lennek

They come from city places like Rockaway and find a new world.

Photo by The Delaware County Times

You have to feed them, wash them and clip them.

Photo by The Delaware County Times

The beef auction, which began in 1980, has attracted an increasing number of exhibitors and buyers each year.

Photos courtesy of Irma and Leo Pierce

Here is Lorinda Pierce Backus, showing a small calf when she was a small girl living on the farm in Mundale. . .

And here she is, considerably more grown up, in the FFA showmanship class in 1987, the year she won top honors.

Photo courtesy of Beatrice G. Bennett

A contestant in the 4-H milking contest in 1949, Lynn Bennett, who then lived in Sidney Center, washes the udder of his cow before starting to milk.

Photo courtesy of The Delaware County Times

If she would only hold her head up! David Davish with his Senior Yearling Holstein.

Photo byThe Delaware County Times

Easy does it!

Photo byThe Delaware County Times

It's a week of work, play, sunshine and socializing.

The Cows That Inspired the Gina-Eva Trophy

Photo by Maggie Murphy, courtesy of *Holstein World*

SHADOWCLIFF R A GINA (EX-97) was shown as a two-year-old at the Delaware County Fair and went on to become All-Time All-American 3-Year-Old. She also holds a lasting place in Holstein history by virtue of inclusion in the Centennial Edition of "Progress of the Breed – the history of U. S. Holsteins" by Richard H. Manfield, published in 1985 by Holstein-Friesian World, Inc. She had been shown in Walton by Carl Hulbert's Catskill Holsteins.

Photo by Maggie Murphy, courtesy of *Holstein World*

A favorite of Dreamstreet's Frank Wood and Robert Harby was HOWARD HOME VALIANT EVA (EX-95) who was undefeated as a 2-year-old and a unanimous All American.

Photo courtesy of Frank Wood

The first winner, in 1998, of the Gina-Eva Trophy was Combination Astre Judy of Pineyvale Farms, owned by the David Post family of Stamford. Left to right in the photo are Astre Judy, Miss Carolee Post, David Post, Bob Harby, Frank Wood, and Mrs. Carl Hulbert.

Photo courtesy of Frank Wood

The trophy itself can be seen at the Cooperative Extension headquarters, Route 10, Hamden.

Photo by Maggie Murphy, courtesy of *Holstein World*

Campbell-Hollow Ultimate Kate, (EX-95), owned in 1983 by Frank Wood's Dreamstreet Holsteins and shown at the Delaware County Fair. Kate went on to be Reserve Grand Champion at the World Dairy Exposition in Madison, Wisconsin and Reserve Senior and Reserve Grand Champion at the Canadian National Exposition.

Photo by The Delaware County Times

Oxen used to be the creatures of choice for much farm work, but then modern times arrived. However, some people still enjoy working with them, and Clem Murphy of Hamden is one. These are Ike and Abe, a pair of Durhams—the original name for what are now called Shorthorns.

Photo by The Delaware County Times

A fine cow is a joy forever, especially if she wins the premium.

Photo by The Delaware County Times

Ann Tyler of Deposit, wearing her heifer's blue ribbon at her waist.

The Farm Trot and On From There

Photo by The Delaware County Times

There was a time, around the end of World War II when horses seemed to be on their way out. *Equus caballus*, which had been alive 60 million years before *Homo sapiens* and had been the beast of burden from prehistoric times into civilization, was about to go out of fashion, especially the draft horse, the old farm plug and puller of wagons. But fickle fashion changed her mind. Today horses, both heavy and light, are not only still with us but also are increasing in numbers and providing sport, pleasure and companionship.

It is easy to picture the County Fair of 1841 with rows of farm carts and buggies tied up to the fence while the whole family ambled over to the race track for the Farm Trot.

Throughout history, horses had been mainly the property of kings, knights, nobles and the very rich, but here in the pioneer farm country, horses had to be for everyone. While more prosperous farmers had finer horses and wagons, all of the families who lived out on the land had to have a horse, and preferably a team. So when it came time for the Fair, out on the track trotted the farm horses whose owners vied in good-natured competition to see who had the fastest and liveliest. In doing this, they were following an impulse which went back thousands of years. Racing is believed to have begun with the first chariots and the first riders.

About the time that the first Fair was being held in Delhi, or a little earlier, archaeologists in Kent, England, were digging up bones that they did not then recognize as fossils of the ancestor of the horse. About 75 million years ago, a prehistoric group called the condylarth family provided the genes for all hooved animals (including zebras!) and over the space of time, a small dog-sized animal emerged.

This was *Eohippus* the "dawn horse," estimated to have lived about 60 million years ago. Its fossil was the one discovered in Kent, but it was not until 1876 when a complete skeleton was unearthed in the United States, that scientists understood what they had. Then, continued study resulted in a realization that *Eohippus* had evolved into *Equus caballus*—the horse as we know it today. The discovery also revealed that North America was the probable birthplace of the horse.

At what period horses were tamed and began to be useful is uncertain, but the Assyrians and Persians are given some credit. Later, in Europe, riding was not well regarded; important people rode in chariots.

However, sometime during the 1500s, riding masters from Italy went to England and a riding school was founded in Spain (from which the Spanish Riding School in Vienna gets its name), and ever since that time, England and horses have been closely linked.

In the 1700s, a kind of racing craze developed in England, which sent its "ambling and pacing" horses to this country. They arrived both before and after the Revolutionary War, thus coinciding with early westward migrations in the Colony. Pacing and trotting races are an essentially American institution and many of the descendants of the rejected British "amblers" and pacers became stars in their own world. In the farming country, people raced whatever kinds of horses were in their stables at whatever pace it was possible for them to go. Consequently, the phrase "Farm Trot" had a fairly interesting background and the term was used until well into the 19th century.

By mid-century, specialized breeding had gotten under way and horses were in great demand. Carriage driving, followed by "equestrian riding" was "the crowning exhibit of the day" in 1856 at the county fair held in Walton under the auspices of the Delaware County Agricultural Society. That is, riding was okay, as long as the rider was a male.

A rather grumpy and chauvinistic columnist for the local newspaper remarked that "the ladies' performance in riding and driving was well done, but not in accordance with our taste. We do not fancy it any more than Mr. Dickinson seemed to fancy Shanghai hens.[1] But there is no accounting for taste in this strange world."

At the 1865 Fair, also held in Walton, 65 horses were entered and there was "a show of matched horses including a 14-span." A letter to B. P. Johnson, secretary of the New York State Agricultural Society, reporting on this Fair, boasts that "the exhibition of horses for numbers, quality, style and action never has been equaled in the county."

By 1887, when the Delaware Valley Agricultural Society began its long existence, there were classes for "stallions, colts and carriage horses."

The Farm Trot had prizes for the "fastest trotting single horse and double team," and for the "fastest walking horse." There was a "go-as-you-please race" of five miles and a Grand Procession with horses stepping out proudly with all of the other animals entered at the Fair. No horse was allowed "in the farm race that has ever trotted for money, excepting in these farm races."

In 1889, all kinds of horses were shown together: Clydesdale, Percheron, Shire and French Draft were the heavies, and there were three lighter breeds not much heard of any more, French Coach, Hambletonian (which became the name of a famous trotting race) and Cleveland Bay.

The Cleveland Bay, a carriage horse, is one of the oldest British breeds. It resulted from crossing huge medieval Chapman horses with Thoroughbreds, and it is known for its prepotency. In the late 1800s, it was being used so widely for crossing, to satisfy the demand for lighter and faster carriage horses, that a Cleveland Bay Horse Society was formed in 1884 to try and save the purity of the breed. A stud book was published, and efforts were continued to bring the horse back from the brink of extinction. More recently, Queen Elizabeth II of England is said to have shown interest in this breed. Still, the Cleveland Bay is now on the endangered list of the American Livestock Breeds Conservancy.

It is possible to foresee, even at this early time, the eventual division of the horse department into three sections: Light Horse Show, Draft Horse Show, and Trotting and Running Races. The races were always separate; the breed judging, driving and equestrian classes developed gradually and increased in number as the "draft horses," "farm chunks" and "general purpose farm horses" slowly faded away.

In 1934, the entire Fair became "The Delaware County Fair and Horse Show." E. F. Runnells of Walton and Harry Hopkins of Lake Delaware were horse show superintendents, and there were 38 classes, which included a great variety of entries: Five-gaited saddle horses, road hacks, harness horses (with appropriate vehicles), draft horses, polo ponies, a tandem class, a jumper's sweepstakes and a Touch-and-Go jumping event.

Jumping, as a show event, at that time was at the most only 70 years old. It is said to have taken place for the first time at a show in Ireland in 1864, but it did not reach international proportions until 1907 in London.[2] It really came into its own only after the Second World War.

With Runnells and Hopkins again the superintendents, 1935 was to be one of the biggest years for horses until well after World War II, and it was the year the department was divided into five sections:

Saddle Horses, 31 classes
Farm and Draft Horses, 26
Harness Horses, 4
Ponies, 10
Jumpers, 11

making a total of 82 classes.

Spotted horses (Appaloosas) evidently were well liked because there were two classes for them. There was also one for Heavy Weight Saddle Horses and a "Saddle Comic Class" in which contestants wore humorous costumes. A riding school/camp class to be entered by an instructor and three pupils was judged both on riding and the manners of the horse.

The harness classes were for a combination of harness and saddle horses—roadsters, taller than 14.2 hands; tandems and "fine harness horses to be shown before an appropriate vehicle," with 50 percent to be earned by the appearance of the vehicle and 50 percent conformation and harness.

The final class was an exhibition of various breeds and types: "Trotting horse, three-gaited, five-gaited, farm chunk, running horse, Indian pony, hackney, jumper, hunter, polo and military mount."

A Horse Pulling Contest took place for the first time in 1937. One of the rules was: "Drivers shall abstain from profanity and shall not use a whip!"

The horse show committee had grown to include Harry Main of the Gerry estate in Delhi as chairman, Frank Lamport of Hobart, John M. Briggs of Deposit and William C. Tyler of Roxbury. The number of classes had declined to 36 divided evenly between Light and Draft Horses.

Two years later, in 1939, Harry Main was joined by Earl E. Jenkins and Johnnie Reside of Margaretville, and Dr. Arthur Sutherland, a veterinarian from Andes. The scheduling pattern remained the same, but the number of classes had continued to fall.

During World War II, and throughout the 40s, the Horse Show remained more or less stabilized with essentially the same number of classes and Harry Main and later William Tyler as chairmen. The committee continued to grow as E. Ogden Bush of Walton, Mr. and Mrs. Floyd Cronk of Grand George, Mrs. Tyler, and several others, joined the group. Premiums rose from a total of $927 in 1944 to $1,024 in 1949.

As the Fair entered the 50s, draft horses had disappeared. Actually, about that time, many people were saying that the day for all horses had passed. Soon it became apparent, though, that horse racing was going to become a major sport and a curious phenomenon presented itself: More and more people were buying horses and riding them for pleasure. The Depression was over and many people had more money. It began to be argued that life was becoming so over-mechanized that some kind of counterbalance was needed. A horse could represent relaxation and leisure. It was also suggested that horses were becoming a new status symbol.

The Light Horse Show

What is a Light Horse Show? It is not a light "horse show." It is a "Light Horse" show. Horse talk has a most interesting history and goes way back. For example, at the show one may hear people speak of "warm-blood" and "cold-blood" horses. This has nothing to do with being warm- or cold-blooded, but rather with where a breed's ancestors came from. In other words, from what part of the world the breed, or "blood," [3] apparently had its beginnings.

So-called "cold-blood" horses are said to have originated in cold, northern regions where the climate varied between chilly and comfortable and where pastures were green and plentiful. Not having to fight for food, they developed a more or less mild temperament; at the same time they were very large and powerful. These are the Heavy Horses.

"Warm-blood" horses, on the other hand, are the result of crossing the heavy, "cold-blood" horses of medieval Europe with "hot-blood" animals from Arabia, Turkey and Spain. These lived in hot, dry countries where food and water was a problem and there was more stress. Consequently, they became more highly strung, but they had tremendous stamina. Familiar breeds such as the Thoroughbred, Standardbred, Morgan Horse, Quarter Horse, Appaloosa, Palomino, in fact, most of the animals at the Light Horse Show are descended from "hot-blood" animals crossed with heavier breeds. They are "warm-bloods;" they are the Light Horses.

There is no doubt that horse shows have a certain cachet, and this may reflect the formal, aristocratic ambiance prevalent throughout the long years of admiration for horses in England. Life is more casual than it used to be, but formality and elegance still characterize much of the Light Horse Show in Walton. In the premium book, a small notation reads: "All contestants must wear appropriate clothing, footwear and headgear for their division and in accordance with American Horse Show Association rules." Horses have been described as proud animals and those who ride them at shows are challenged to present both horse and rider in a superlative way.

Considering the price to be paid for show saddles, bridles and other gear, plus boots, breeches, formal jacket and other garments, to say nothing of the cost of buying and keeping the horse, showing is expensive. Yet a growing number of families in modest circumstances are willing to give up other pleasures to keep and show horses —as the steady growth of the show in Walton testifies. Each year the bleachers fill up earlier and the number of spectators increases.

The event also has come to be known as The Silver Horse Show because, along with the ribbons, it has a tradition for presenting silver—and sometimes crystal—trophies.

The horse show program book lists these divisions:

- Hunter Division Horse and Pony
- Western Division
- Gymkhana Division
- Saddle Seat, Driving and Lite-shod Walking Horse Division.

These may include classes such as:

- Equitation classes which are "judged on seat and hands, general knowledge and ability to control the mount..."
- Classes for very young riders, eight or ten years old and under.

- Competition for Best Rider in Delaware County.

The Best Rider competition began with the saddle seat division in 1953 and subsequently was enlarged to include Western and hunt seat. The Western, or stock seat, competition began in 1958 and the hunt seat in 1964. To compete, a rider must be a year-round resident of Delaware County and not older than twenty-one. The first Best Rider trophy in the Saddle Seat Division was won by Linda Landers, in the Western Division by Ginger Fowlston and in the Hunt Seat Division by Ilene Cochrane.

Each division attracts a certain type of horse rather than a specific breed. In the Western division, for example, a rider could be seated on a Quarter Horse, Morgan, Paint, Appaloosa, a Western-type pony, or any stock-type animal, that is, a horse that would be working with cattle.

A hunter-type horse would be longer-strided for galloping cross-country and jumping—for example, a Thoroughbred, Quarter Horse, mixed breed, Appaloosa or Pinto. Actually, a heavy hunter could be a cross between a Thoroughbred and a draft breed.

A saddle seat horse would have an alert, "up" stance, its neck arched, and higher knee action. A Morgan horse, Arabian, or American Saddlebred (they are sometimes called "the peacocks") are often seen here.

Very much admired are the Plantation Walking Horses which first appeared in Walton in 1955. After a number of years the premium book stops listing them, but in 1985 they are back. According to show secretary Lois Brandow, Walton is one of the few fairs that attracts Walking Horses. Many come from as far away as Albany, Syracuse and some places in New Jersey.

Almost any type of horse can be hitched to a wagon or cart in the driving section. These horses are judged on "manners, performance and quality." For the past few years, the driving section has been the fastest-growing division. Walton attracts two types of driving horses:

- American Driving Society horses which are used as much for cross-country driving as they are for show. They are naturally shod and the appropriateness of the turnout to the two- or four-wheeled cart is important.
- Show Pleasure Cart horses wear a much lighter harness. They pull two-wheeled carts which have pneumatic tires; their speed and action as well as manners and elegance are important.

The Gymkhana—games on horseback—like many horsy things, had its modern origin in England where all through the 19th century it was part of a typical local festival. In Walton it has grown from a small program to a popular event scheduled for the entire evening. In 2000, there were 20 Gymkhana events including pole bending, barrel racing and the old Ride-a-Buck in which a bareback rider attempts to sit on a dollar bill and keep it there. It functions under the American Horse Show Association and American Quarter Horse Association rules, and it attracts contestants from as far south as Middletown and north to Ithaca and Albany. Gymkhana events, usually fast and furious, are ridden in Western-type clothing and tack.

Organizing the Horse Show is a formidable enterprise. Listed in the 2000 premium book was the committee, consisting of John Johnson of Unadilla and Larry Hillis of Hobart, cochairpersons; Lois C. Brandow of Franklin, show secretary; Kim Hyzer and Sylvia Fancher, computer secretaries, and Kay Crane, Roberta Austin, Mary Hafele, Donna Hyatt, Kathy Wheeler and Michele Hyatt, assistant secretaries. But that was just a drop in the bucket, according to Lois, who has worked with the show for many years. There are many, many more volunteers who work long and hard to make the show successful.

Not a volunteer, but one of the most important personalities, is the announcer—a pleasant voice at the mike and a champion expediter. Since 1984 he has been Kenn Marash of Harford Mill, New York, who is also the State Fair horse show announcer. His predecessor was Jason Gates of Cheshire, Conn., who served for a number of years and also did the State Fair.

A glance at the premium books shows that in 1955, Mrs. George Wilson of Delhi was chairman of the horse department and Patricia Schreiber of Walton was secretary. The rules that year contain an interesting note reminiscent of the earliest days, one which was carried for some years: "Trained race horses barred from this department." Eight classes were for Delaware County residents only and there were 22 open classes including those for Road Hacks, Three-Gaited Saddle Horses, Western, Walking Horses, Jumpers and so forth.

By 1957 a set of divisions had been created. With Mrs. Wilson still chairman, there was a Three-Gaited Division, a Western Division, a Walking Horse Division and a Jumper Division in addition to nine county-only classes and three open classes for "obedience" and "seat and hands."

In 1963 John "Bud" Mckown of Franklin became the chairman and he served for the rest of that decade. Mrs. Edmund (Lois) Brandow was secretary for a year, followed by Mrs. McKown and Mrs. Lorin R. Skiff and Mrs. David Doane, both from Walton. For some reason the Divisions had disappeared and once again the events were mostly mingled, a pattern that lasted for about 10 years.

With the 70s, a succession of chairmen took over including Richard Nelson of Downsville and Lorin R. Skiff and William Oles, both of Walton. Lois Brandow was back as secretary in 1974 for a year, followed by Mrs. Skiff and Mrs. Lois W. Ray of Delhi who served until 1979.

By 1974 the divisions had reappeared. They were: 1) English and Pony Division, 2) Harness Pony/Hunter Horse, combining the driving and jumping, and 3) Western and Gymkhana. By 1979 these had been refined to: English and Driving, Hunter and Pony, and Western/Gymkhana. By the end of the 80s, these titles had become more or less what they are today.

Arnold Froehlich of Walton became chairman in 1978 and served until he retired in 2000 and was succeeded by John Johnson and Larry Hillis. The three-day show in 2000 was dedicated to Arnold who "contributed more than 20 years of hard work and dedication as Chairman of the Open Horse Show. His efforts resulted in establishing this horse show as one of the top county horse shows in New York State." Arnold and his late wife, Doris, have been greatly esteemed and much admired.

Mrs. Marion Budine of Walton became secretary in 1980, Lois Brandow returned as assistant secretary in 1986 and took over as secretary in the early 90s. She taught judging locally and regionally for many years, served as horse specialist for the Cornell Cooperative Extension and has been a leader of the 4-H Franklin Lite Brigade.

Today, as she looks back at nearly forty years with the horse show, Lois Brandow can provide a bird's-eye view. In 1973, there were 82 classes in a three-day show; there are now 142. Years ago there was a barn next to the grandstand which was used for the horses, and the horse show ring was opposite. Horses' stalls were under tents like everything else at the Fair, and exhibitors had to go down to the river for water. Now the spacious new horse barns built in 1981 and 1982 are a source of pride.

A recent development is inclusion of the New York State Horse Council's Millennium Horse Medallion Program which offers special awards in all of the divisions. The Horse Council is devoted to promoting the horse industry in New York State. In a sense, things have come full circle. Once again horses are not just for pleasure; they are a business.

And the latest: The Horse Show is on the Internet.

The Draft Horse Show

It had humble beginnings. Ann and Vincent Warner and Carol and Robert LaTourette were the guiding spirits, but they had run into somewhat uneasy sanction on the part of the fair board. The day of the show, in 1974, it rained. Stables were not available, so the horses were tied to trucks in a vacant parking lot. The rain streamed down on the animals, all grouped together.

However, the response to that first effort was so enthusiastic that everyone soon decided that yes, after 25 years, draft horses were on their way back. Soon the show began to be held in the main horse show ring and Monday was designated as Draft Horse Day.

So many draft horses had been slaughtered in the period after World War II that they were not just "endangered," they were nearly extinct. What saved them? Some credit the lifestyle of the Amish and Mennonites in Pennsylvania. These people rejected tractors and new farm technology in favor of the old ways, and they loved horses. Also, while many breeders had sold their draft horses during and after the War, a number of mid-western farmers refused to let go. They were the means of preserving precious and irreplaceable bloodlines which went back many centuries.

The Belgians are perhaps the most familiar breed at the Draft Horse Show and a favorite of the late Jack Briggs and of Bill and Marge Rockefeller of Delhi who have shown them for years. Belgians, massive, red-sorrel animals are descendants of the Great Black Horse of Flanders. Over the centuries their color was greatly modified. In his book, *The Illustrated Horse*, writer Tom Howard remarks that breeders in Belgium did not cross their mares with Arabian, English or French stallions as other countries often did.[4]

The late Wendell Spear of Walton, who, after retiring as principal of the Walton Central School, turned his attention to horses, also was a Belgian enthusiast.

During the 80s and 90s, spectators at the draft horse show watched as the shiny, black, exquisitely trained Percherons fitted with silver trappings pulled the white Skyline Farms wagon driven by Jack Mulligan of Earlville. Often the only entry in the six-horse hitch, the Percherons dazzled onlookers with their precision and impeccable training. Percherons were, in cart horse days, the best-known breed. They originated in the La Perche region of France and were interbred with Arabians which accounts for their liveliness. Joan of Arc is supposed to have ridden a Percheron.

A third great, heavy horse which appears at the Fair is the Clydesdale, which comes from Scotland. Clydesdales are famous for the "feathers" on their legs, as well as for being the famous Budweiser Beer Horses. Although extremely powerful, they are known for possessing a quiet disposition.

The Shire Horse, whose origin goes way beyond any existing records, is "the national draft horse of England" and gets its name from the two "shires," or counties, where it was bred. They, too, have shaggy hair growing below their knees and hocks.

The year after that first successful event, the Draft Horse Show was listed in the premium book with Robert McNee of DeLancey as superintendent. There were three sections for registered horses: Belgians, Percherons and Clydesdales; there were two classes for "grades" and six for hitches—singles and pairs, a ladies' cart event, a driving competition for teams, a three-horse (unicorn) hitch and a four-horse hitch.

By 1976, Shires had been added to the list of registered horses, and in 1979 eleven exhibitors came from as many communities in New York State—five from Delaware County. There were 46 horses representing the four breeds; they competed in 19 classes.

Also in 1976 the administrative work began to be shared by Vince Warner of Masonville, Robert

LaTourette of Sidney Center and Jack Briggs of Deposit. They functioned as a trio until the untimely death of Jack Briggs in 1984.

A tribute to his memory was included in the premium book for that year. A lover of horses all his life, he had bought his first Belgian three years after graduation from Deposit High School in 1933. He had attended Cornell University and in 1945 had gone to Cornell as Superintendent of Draft Horses and Farm Manager with the Department of Animal Science. He had been a draft horse judge at many fairs and he had bred and shown many a champion.

After her husband's death, Faye Briggs became a Draft Horse Show committee member, joining Robert LaTourette. Vince Warner and Marjorie Rockefeller of Delhi were the cochairmen. Mrs. Briggs served until 1987. Gladys Crane of Walton joined the committee in that year.

The six-horse hitch was instituted in 1985 and about this time a "ride a draft horse" event began to be held. Marge Rockefeller became chairman in 1989.

Draft horses are shown in two ways: In Halter Classes, which concentrate on conformation and performance, and Hitch Classes, which involve various pairings such as six, four and three-horse hitches, pairs, tandem, driving carts, and single horses. Included are classes for horse and pony Farm Teams and for Farm Carts. Through the years the Draft Horse Show has attracted nationally-recognized judges such as Robert Dunton of Michigan, a director of the National Belgian Association, and Corbly Orndorff, whose Pennsylvania horse farm has produced some of the best Belgians in this country.

In recent years an "Old-Tyme Draft Horse Show" has been added, its purpose being that participants "have fun." Adults or young people may enter a horse, pony, or mule pulling the required wagon or cart and using heavy harness. Trophies have been established—the Bert Sines Memorial Prize and the Millard and Florence Kilpatrick Memorial Prize.

Horse Races

Harness racing took place even before the Revolutionary War when the main streets in Maryland and Virginia towns are said to have served as a kind of race course. Not surprisingly a race track was one of the first things to be built at the early Delaware County fairs.

When the Delhi Agricultural and Mechanics Society was founded in 1862 in Delhi, the organizers immediately built a track.

Likewise in Walton in 1887. "The fairgrounds have been surveyed and a one-half mile track staked out," wrote an enthusiastic reporter. Subsequently, harness racing as well as running races were a tradition at the Fair in Walton until the 60s when so many attractions were crowding the area inside of the track that the judges in the stand at the top of the track couldn't see the quarter posts. So the races had to go. Many people still miss them. Perhaps with good reason, for horse racing seems to have been part of becoming civilized.

There was racing at the Olympic Games in ancient Greece; horses were raced in all of the Arab countries. The early Christians excommunicated chariot racers because the sport had been associated with pagan festivals, but apparently this did not have much effect because in ancient Rome it was one of the most popular excitements.[5]

Racing disappeared during the so-called "dark ages"—as did fairs—but in Britain, about 1170, there are reports of horse fairs where, sometimes, races took place. As time went on, racing became well established—in Britain—but not on the Continent. About the time of Henry VIII, the British began to think seriously about the blood lines of their "running horses" or "Galloways," as they were called, and they started to breed their best runners. When James VI of Scotland became James I of England, he became the first monarch to create a royal stud. After the Civil War, Oliver Cromwell banned racing in England, but he did preserve some of the royal horses. The Jockey Club "Rules of Racing" became universally applied in 1870.

As the "running horses" became more popular in England, the "amblers" and "pacers," who were really very good, fell out of favor, and thousands were shipped to America. Here, harness racing took hold as it never had in Europe, and, building on the bloodlines of many of the immigrants, a succession of extraordinary trotters and pacers was produced for the races in America.

Unlike flat racing, "the sport of kings," which generally took place near urban centers, harness racing thrived in agricultural areas, in the rural towns and villages. Hence, the Fair in Walton was a natural center for the sport, and a writer in 1933 declared that horses came "from every part of the United States to run at the Walton Fair."

In the 1930s, E. J. Kellam of Hancock was in charge of harness racing; each year he forecast bigger and better things, but at that time the outlook for horses was not very good. In 1935 The Walton Reporter remarked that "harness racing...is fast losing its old appeal." It lasted, however, for three more decades. At the same time, running races were being held under the chairmanship of Arthur Brundage of Margaretville. They lasted only a couple of years.

In 1948 Lloyd "Doc" Segar was appointed racing secretary and that year saw the introduction of a new automatic starting gate. At the same time the three-heat-per-race system ended in favor of two heats run with a $450 purse split between the winners of each heat.

The end of harness racing came in 1964.

Photo courtesy of Veda Thomson

The Light Horse Show in progress, 1989.

Photo by The Delaware County Times

Bill Rockefeller of Delhi, with his Belgians in a three-horse hitch, comes round the curve during a draft horse show.

Photo by The Delaware County Times

Draft horses are shown in halter and hitch classes. This Shire Horse will be judged on points to determine how close the horse comes to a standard which defines the ideal animal in its breed. In the background is the old announcer's platform, which, for many years, stood opposite the grandstand.

Walton Historical Society Photo

In 1948, the horse show ring was alongside the track, opposite the grandstand, and cattle were housed in the tents to be seen in the background.

Walton Historical Society Photo

An equitation class being judged at the Fair in 1948.

Photo courtesy of Lois Brandow

For many years the horses, like almost everything at the Fair, had a tent over them. Here, under the tent in 1973, are Kim and Deb Brandow and Chris Robinson with their horses.

Photo by The Delaware County Times

A classic shot of a classic horse show event. Probably taken about 1995.

Photo courtesy of Lois Brandow

Western-style riding became increasingly popular as the years passed. Deb Brandow is riding Mr. Buck, a registered quarter horse, in a pleasure class sometime during the 1970s. In the background is the old horse barn which was subsequently replaced by the 4-H barn. The Light Horse Show had 82 classes in 1973; in 2000, there were 142.

Photo courtesy of Mary and Joe Hafele

Two of the most admired breeds of horses at the shows are Morgan Horses and American Saddlebreds. Liza Hafele, left, is riding Diamon' Lil, an American Saddlebred; Molly Hafele's mount is Delview's Chasely Ruben, a registered Morgan.

Photo by The Delaware County Times

A dream come true: To be astride a beautiful horse, holding a blue ribbon!

Photo by Robert N. Lennek

The younger, the better.

Photo by Robert N. Lennek

So pleased was the Horse Department with its refurbished ring in 2000, that fair board members were invited to come and try it. Lois Brandow, in the white outfit, is making sure that each member is properly mounted on horses provided by various exhibitors.

Photo by The Delaware County Times

"Ride a draft horse" is a fairly recent and popular class.

Photo by Robert N. Lennek

Don't be surprised to meet a cart like this on your favorite rural road one of these days.

Photo by The Delaware County Times

Instead of disappearing from the scene, draft horses in this country made an astounding recovery and emerged with renewed prestige. The first Draft Horse Show was held at the Delaware County Fair in 1974, and it has grown in popularity ever since. A new draft horse barn was built in 1986.

Photo by The Delaware County Times

A uniform appearance in size, color and conformation is desirable for horses in a team or hitch.

Photo by The Oneonta Daily Star

These splendid Percherons from Skyline Farms, Earlville, have dazzled spectators at the Fair for at least a decade. Jack Mulligan is the driver.

Walton Historical Society Photo, courtesy of the E.B. Guild family

Pacers at the Fair, about 1941.

Those Ancient and Lovable Goats

Photo by Robert N. Lennek

Visitors to the Fair are always intrigued when they enter the goat section to see an array of pointed ears and whimsical, mischievous faces turned toward them as the animals rest their forefeet high upon their dividing partitions to survey the scene.

Goats, someone has said, regard themselves as folks; it is one of their most lovable qualities. In terms of history, goats are ancient, having been mentioned in Sumerian cuneiform tablets as early as 14 centuries before Christ.[1]

Dairy goats have been seen at the Fair in sometimes large, sometimes small numbers since the early 1900s. The premium book for 1909 lists, at the end of the Sheep Department, seven classes for goats. They were of no particular breed or description for the simple reason that purebred goats were only beginning to be imported. The premiums were $2 and $3, depending upon age, and $5 for the best pen.

At the 1916 Fair, which was a big one any way you look at it, there were four different sections for goats: *Angora, "Thoroughbred" Milch Goats, and "Grade" Milch Goats*, all of which had to be registered with the American Milch Goat Record Association, and *Native Milch Goats.*

The 4-H began to have classes for goats about 1948. Also in that year the Catskill Dairy Goat Association became organized, but it wasn't until 1950 that the first official Dairy Goat Show was held.

What happened was that a number of goat enthusiasts in Delaware, Otsego, Schoharie and Broome counties got together to form the CDGA and they chose the Delaware County Fair as the place for an official dairy goat show. Among these breeders were William Ostrander of Sidney, owner of the *Stoneydel* herd of Alpines, and Mr. and Mrs. Howard Dorland of Stamford who were breeding Nubians. Mr. Ostrander was among those instrumental in getting the Oberhasli breed accepted by the American Dairy Goat Association. A third breeder involved was Elizabeth Nicholds, a writer who had given up a lucrative job in the city to buy a farm in nearby Otsego County and raise goats. *Thunder Hill Farm,* her book about her experiences, became immensely popular and later a lavishly illustrated article which she wrote for *The National Geographic Magazine* made her something of a celebrity. With these people in the lead, the Catskill Dairy Goat Association grew and prospered.

There were 63 dairy goats at that first show, on August 17, 1950, more than would appear at the 1950 State Fair. Myra Smith of Walton was the show secretary, and the entry fee was 50 cents. The show was open to any resident of Delaware County, as well as any member of the Catskill Dairy Goat Association from any county. Crawford Brothers furnished free grain; the Fair Board provided ribbons, 100 of them, bearing the words *Walton Fair* and *Catskill Dairy Goat Association.* It also provided the clean and draft-free housing.

In those days there were only four main breeds of dairy goats: French and Swiss Alpines, both from Switzerland; Nubians from North Africa; the pure white Saanens and the brown-coated, white-booted, striped-faced Toggenburgs, both from Switzerland. Now there are six main breeds: French Alpines, Oberhaslis (which used to be Swiss Alpines), Toggenburgs, Nubians, Saanens, and LaManchas, a breed created in this country.

Toggenburgs are said to be the oldest registered breed of any animal in the world; their herd book goes back to the 1600s in Switzerland. They were apparently the first purebreds to be imported (in 1893) and have been rather an elite goat. The poet, Carl Sandburg and his wife, Paula, had a well-known herd of Toggenburgs.[2]

Saanens, a quite large goat, arrived here in 1904; they are supposed to be pure white but occasionally a patterned animal, the result of a recessive gene, made its appearance, and these now can be shown as "Sables."[3]

Swiss Alpines, a reddish-bay goat with black legs and a black stripe down its face and back, began to come to this country about 1906, but for some years they were not raised as purebreds. In 1936, Dr. H. O. Pence of Kansas City, Missouri, imported a few purebred Swiss Alpines and the breed grew in numbers. Their name was changed to Oberhasli in 1978.[4]

Nubians, those fantastically gregarious and mischievous Roman-nosed, floppy-eared goats, presumably originated in Africa and India. Today's Nubians are the result of crossing English goats with these foreigners. They began to come to America in 1909.[5]

French Alpines are the latest comers, the first having arrived in 1920,[6] and they probably have the most variety of colors. They are also probably the most popular dairy goats at shows today.

One of the most recent dairy-goat breeds, the LaMancha, was developed in America during the 1930s by crossing short-eared domestic grades with Alpines and Nubians.[7] There is one standard reaction to meeting a LaMancha for the first time: "*They have no ears!!*" But those who breed them boast of their productivity and amiable dispositions. In 2000, there were 48 LaManchas at the Delaware County Fair.

Among those outside of the dairy-goat category are Angora goats, which produce mohair, and the African Pygmy and Nigerian Dwarf, which are growing in popularity. From the standpoint of supply there should never be a shortage of goats at the Fair although there may well be a shortage of people who raise them.

As often happens to people-groups, the Catskill Dairy Goat Association in the late 1950s went into a decline, and there were very few goats exhibited in Walton for well over a decade. Then, early in the 70s, the scene changed. A "back-to-the-country" movement started to develop, and an interest in alternative farming began to arise. These developments seem to have spurred a renewed interest in goats. Not surprisingly, at that point they became expensive— $300 for a nice purebred in the mid to late 70s was not unusual. The Catskill Dairy Goat Association was reorganized, and in 1979 the fair board was asked to reactivate the open goat show. H. Gill Brown, a licensed judge and breeder from Pennsylvania, officiated. It was a small show, but it represented a resurgence.

Early in the 1980s, Allen Bitter, a Delaware County Cooperative Extension agent, initiated a new, "sanctioned" one-day open goat show which flourished for some years. Often as many as 300 goats came from far and wide since this show was on the American Goat Association "circuit." Don Byington was the goat department superintendent between 1980 and 1985. Marge Warner, a Nubian breeder and long-time 4-H leader, became superintendent in 1986. She is among a number of adult committee members who participate regularly in the annual Farmer's Museum Junior Livestock Show in Cooperstown.

Linda Smith, who for many years has been raising the award-winning *Sherman Hill* Alpines at her farm in Franklin, became secretary of the dairy goat department in 1979; in 2000 she shared the title of superintendent with Marge Warner for the open show. Linda was one of the original members, with William Ostrander and Gordon Tyler, of the Dairy Goat Club in 1978. The organizers held a series of "interest meetings," either in people's homes or at the Franklin Firehall, and invited guest speakers or held workshops. Sometimes as many as 200 enthusiasts would show up for a workshop. In 1984 some of these people were instrumental in setting up a national convention. But by the beginning of the 90s, other commitments began to chip away at the devotion to goats; once again interest declined and the club was dissolved about 1995.

Linda Smith also was an early leader of the Buckram 4-H Club which, as 2001 came into sight, had 18 members. The Buckrams were responsible, in 1988, for building the milking room in the dairy goat barn. The year before, the 4-Hers had worked on a quilt section which they presented to Ms. Smith. She was in the goat barn early Tuesday morning during the Fair of 2000 to oversee the 19 4-H classes. There were milkers, meat goats and "family" groups, showmanship classes and one for pigmy goats. The Cloverbuds, showed young animals in a showmanship class.

These events were followed by the one-day open goat show for female purebred or recorded grades. All of the entries, however, had to remain on the fairgrounds for the entire week. This stipulation has been the source of some displeasure on the part of the goat people since 1996, prior to which the goats came just for the day. In 2001 this problem seems to have been solved with the scheduling of the open goat show on Sunday before the opening day.

Today, with small farmers showing an increasing interest in alternative farming, meat goats are growing in popularity (they are eligible for government subsidies). Raising meat goats is less labor-intensive than milking a herd twice a day, and there is no problem of what to do with the milk. *Sherman Hill* is one of the few commercial cheese makers in New York State. But raising goats for profit seems not to be the only reason for keeping them. "People are buying goats just to enjoy them" insist some of the breeders who exhibit at the Fair, and, besides, as many have discovered, goats are good companions.

Photo by The Delaware County Times

The 4-H has been showing goats for nearly 60 years.

Photo by The Delaware County Times

The open goat show attracts exhibitors from a wide area and goats of many breeds and colors.

Photo by The Delaware County Times

Lisa Roland will probably remember this moment for the rest of her life.

Photo by Robert N. Lennek

Look how friendly they are, mom!

Photo by The Delaware County Times

A black Nubian makes such a good companion!

Photo by The Delaware County Times

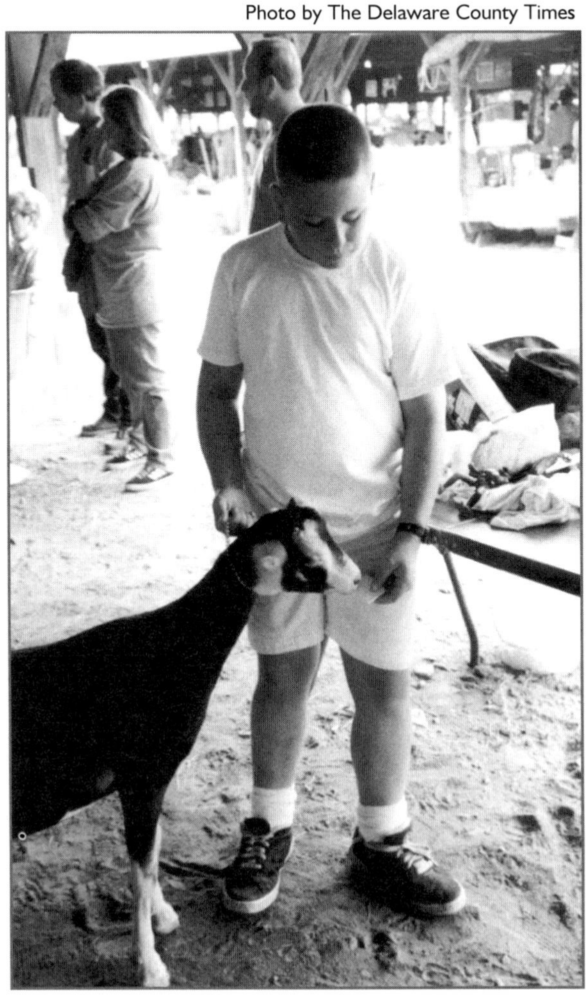

LaMancha goats are a recent dairy-goat breed, having been developed in this country during the 1930s. Some may regret the absence of the standard, long goat ears, but the animals are said to have a delightfully amiable disposition.

Photo by Robert N. Lennek

Goats have to be clipped before they are ready to be shown. A stanchion like this helps.

From the Cornell Cooperative Extension Archives

Sherman Hill Alpines of Franklin have been among the outstanding exhibitors for some years. Here is Morgan George with Pandora in 1994 when she won the Master Showmanship plaque. It is a rotating award, permanently housed at 4-H headquarters in Hamden.

Photo by The Delaware County Times

There's a little bit of stress involved for everyone at a goat show, but there's always another year!

Photo by Robert N. Lennek

Really, I have never seen anything quite like you!

Photo by The Delaware County Times

I just love to watch all the funny people!

Our Scottish Heritage

Kate Henderson photo

It has been said that when Scottish immigrants came upon Delaware County in the late 18th and early 19th centuries, they were reminded of the hills of home and stayed. No doubt about it, there had to be sheep. One historian has insisted that at one point there were more sheep in the county than people.

It's hard to tell what kind of sheep were to be found at the first county fair in Delhi in 1841, since, as with other animal species, many breeds so well known today had not yet arrived. The Horned Dorsets, for instance, that are on so many farms now, did not appear on the East Coast until after 1885. An Englishman exhibited them at the American Fat Stock Show in Chicago. But the Tunis, which came from North Africa and was a favorite of Thomas Jefferson, had come as early as 1800. Both breeds are familiar to fairgoers today.

The Delaware County Agricultural Society, which had organized that first county fair, held several in Walton, including fairs in 1856, 1865 and 1869.

At all three, sheep were apparently very much in evidence. In 1856, according to *The Walton Blade*, there were winnings among Southdown lambs, Saxon and Merino ewes, and Cheviot, Saxon and Bakewell "bucks." Familiar surnames occur among the premium winners: A. Tweedey, (note spelling), B. J. and Russell Bassett, and Charles Edgerton.

The report of the 1865 Fair noted that "The sheep presented included 90 animals. Among them were several imported sheep and their immediate descendants. The competition among the long wools was very strong as a large number of the animals were of superior merit. The show of middling and fine wools was good."

At the 1865 Fair, sheep could be entered in three ways: Purebred, crossbred or fat sheep. Exhibitors of purebred sheep—Cheviot, Merino, Leicester, Saxon, Southdown and Oxfordshire—were asked to produce the animals' pedigrees. The fat sheep and lambs were part of a class called "Fat Cattle and Sheep," and it included hogs.

By the time the Great Walton Fair was born in 1887, Archibald Tweedie was presiding over two classes. One was for purebreds, mostly from English stock: Oxfordshires, Southdowns, Leicesters, Shropshires, Hampshires, Cotswolds, Merinos, "or any other Pure Bloods." The second class was for mixed breeds.

As the years passed, sheep entries rose and fell at the Fair as the cycles of popularity and disinterest came and went, and as the price of wool went up and down. But always there have been at least a few who heeded the advice of an early sage who declared that "the wise farmer keeps sheep," and for some people a farm has not been complete without at least a few.

Early in the 1900s, Andrew Easson of Mundale was a well-known exhibitor. He took his sheep to the Fair either by loading them on a wagon, or you may picture him walking behind them as they trotted the seven miles or so down East Brook Road.

Robert Kilpatrick was in charge of the department from 1903 to about 1910.

At the extraordinary Fair of 1916, often mentioned elsewhere in this book, the sheep department, under the direction of V. A. Wolf, listed 15 breeds: Cheviots, Southdowns, Oxforddowns, Shropshires, Cotswolds, Leicesters, Merinos, Rambouillets, Horned Dorsets, Lincolns, Suffolks, Delaines, Hampshiredowns, Tunis and Sabledowns.

In the 20s, Jay Cook, whose family lived along West Brook, entered the Fair scene by showing chickens, but he shortly outgrew that activity and worked his way up to three railroad cars of sheep, swine and goats. One year, Mr. Cook is said to have taken more than 30 sheep to the Fair. In 1986, when he was ninety-one years old, Mr. Cook visited the Centennial Fair.

Many of the sheep shown during the 1920s seem to have been the fine-wooled, curly-horned Merinos, together with Leicesters and Scottish Highlanders.

One of the most active exhibitors during the 30s was Robert Munn, who, in later years, was the town highway superintendent for Walton. He had a large flock of Cheviots which he showed at many fairs around the state. His sheep were frequent blue-ribbon winners at the State Fair, and one of his rams went on to be shown at the livestock show in Chicago.

The 30s were a peak time for sheep for a variety of reasons, including the growth of the 4-H. The 4-H sheep show accounted for between 50 and 75 sheep a year during that period. Sheep were housed at the end of the cattle tent, making one long livestock area. Before that they had been placed in a two-story barn to the right of the lower driveway into the fairground.

By the end of the 30s, dairy farms were getting bigger and more important to the economy, and the day for synthetic fabrics was dawning. The 40s and 50s were a down period for sheep; they all but disappeared at the Fair.

But then, in the 60s, Paul Miller of Mundale, a descendant of Andrew Easson, and his family began to be exhibitors, as well as the Sulgar, Nichols and Whiteside families. As with the goat and swine departments, the number of entries from year to year has largely depended upon whether there are many or few families with youngsters interested in showing.

In the early 80s, John Campbell of Franklin became department superintendent followed in 1985 by Phyllis Trask Schlippel who had grown up on her family's Reyburn Road farm. A trio, Kate Henderson, Nancy LaTourette, and Pat Winters headed the department for several years during the early nineties, after which Mrs. Henderson and Mrs. LaTourette continued to organize the sheep shows.

In the cattle barn, the sheep section was the first to get a concrete floor—and it all started because of the sheeps' ears. About 1997, after the formation of the Wild and Woolly 4-H Club, the number of sheep was increasing after some years of sharp and discouraging decline. The pens were getting old, however, and the ear tags of the inhabitants were getting caught in the wire fencing material. The department asked the fair board to remove the mesh and replace it with all-wood pens, like the goats now had, and in the process of doing this, the board also decided to put in a concrete floor.

They did the same for the swine, which were at the other end of the same barn, a year or so later.

The Wild and Woolly Sheep Club was organized in 1996 by co-leaders Kate Henderson and Greg Graupman in an effort to change one of those "down" cycles that every department at the Fair has experienced from time to time.

The number of sheep exhibited by 4-Hers had fluctuated between a high of 96 in 1983 to a low of 22 in 1995. From 1979 to 1986 the pens were well filled, but then participation fell off. In 1987 the number had dropped to 23, remained in the 30s for several years, rose to 48 in 1994 and dropped again to 22 in 1995.

Wild and Woolly started with just a handful of members, and it has grown into a group of 30 boys and girls, ages five through eighteen, from all over Delaware County.

The year after the club was formed, the leaders organized a "Sheep Olympics," which now takes place at the Fair on Thursday nights. Each contestant runs a sheep through an obstacle course, which is timed and might include a walk over a 4 x 8 sheet of plywood, a jump over a small pool of water, a gambol through a hoop... Everyone receives a prize, which might be a feed bucket, a hoof trimmer, a lead line or some other useful item.

The club sends wool to a firm in Michigan which makes batting. When it comes back, the youngsters create quilts, which are then sent to the Lakota Indian Reservation in Pine Ridge, South Dakota. They have made something like 13 quilts a year for the past several years, choosing their fabric, cutting the pieces, sewing them together and then tying the quilts. They are not inexpensive—each one has cost approximately $100 by the time the expenses are added up.

Fourteen breeds were shown at the open show in 2000, including North Country Cheviots and Southdowns, according to Kate Henderson. She and her husband, Ralph, raise Horned Dorsets, Tunis and Romneys at their Sheepy Valley Farm in Dryden Brook.

The Delaware County hills may remind Scotsmen of home, but sheep raising is far more profitable farther west, says Mrs. Henderson. Here in the East, a dual-purpose breed like Dorsets are apt to attract more attention than wool breeds like Rambouillets or Corriedales. The 19th century interest in long-haired, fine-wooled sheep is about gone, although there are a few farms specializing in wool for spinning. They raise breeds such as Merinos or the very ancient, four-horned Jacob Sheep.

For local sheep farmers, meat at present is generally more important than wool, and there are a few farms in the county which stock the new American breed, Katahdins, or "hair sheep." These are great for meat producers, since, having no heavy wool, they simplify things considerably.

For those who depend upon selling the results of the annual shearing, however, times have been tough for several years. Generally these people take their wool to a cooperative, but the market was so poor in 2000 that there was no wool pool at all. Sheep raisers are optimistic, however, that once again, as in past cycles, conditions will improve.

Photo courtesy of Kate Henderson

Horned Dorset Ram

Photo courtesy of Kate Henderson

Many kinds of sheep are shown at the Delaware County Fair. These are Romney ewes.

Photo courtesy of Kate Henderson

Dorsets can be seen on farms wherever you look in Delaware County. The wool from black sheep is baled separately and is often used for specialty items.

Photo courtesy of Kate Henderson

The big ram is a Horned Dorset named Sir William, and he comes from a farm in Virginia. Sires like this guarantee quality entries in the sheep department.

Photo by The Delaware County Times

Over the years, exhibitors who work on the committees, as well as those on the fair board, have put time, energy and money into upgrading the barns and stalls and supplying facilities like this wash rack at the sheep barn.

Shearing a Sheep: one, two, three

Photo by The Delaware County Times

First you sit her on her rump.

Photo by The Delaware County Times

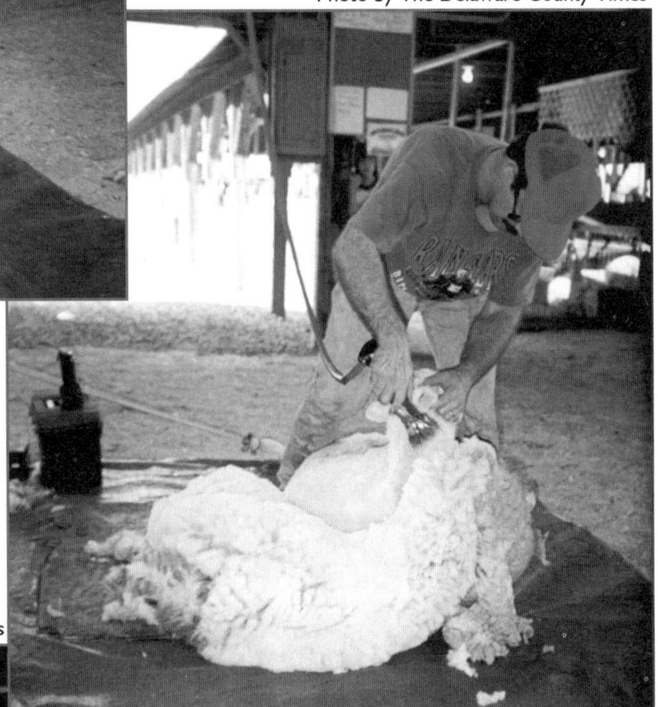

Then you clip.

Photo by The Delaware County Times

Then, when you are finished, you have a cool ewe—and a fluffy pile of fleece.

Every Farmer Had Pigs

Like horses, sheep and cows, most of the pigs that came to America in the early days were of English stock. Yorkshire, Hampshire, Berkshire and Hereford are all counties in England whose barbarous invaders had domesticated the boars of Northern Europe and whose later inhabitants considered them a feast. In the colonies, almost everyone who lived in the country raised a few pigs.

Sometimes these porkers were called Pigs. Sometimes they were Hogs, and when they went to a country fair, they became Swine.

To those who have grown up tending them, the little baby things with the curly tails, as well as the round, fat, sows and boars, are probably best known as pigs, but whatever you call them, pigs, hogs, or swine, today's gourmets, like the people of "olde England," still consider them fun to own and a feast to eat.

At the 4-H livestock auction at the Delaware County Fair, the bidding is fast and keen.

At the county fair of 1856, which was held in Walton under the auspices of the Delaware County Agricultural Society of Delhi, an exhibitor identified as H. Levally won a premium of $3 for having the best boar at least one year old, and C. St. John received $3 for the best sow with four or more pigs. R. Mickelham won $2 for second-place boar and H. Levally had the second-place sow. No breeds were indicated. A. O. Christman was superintendent of swine, sheep and poultry. Hay and grain were supplied by the management.

A decade later, in 1865 and 1869, at fairs also held by the

county group in Walton, swine and poultry were grouped together, with Stephen Berray, Jr., of Walton as superintendent.

John Smith of Walton was one of the judges, together with James Frazier of Delhi, Charles Mabie of West Meredith, G. W. Wheat of Croton (Treadwell), and A. Cable of Deposit. It was very early in the days of showing pure-bred animals, and exhibitors of Essex and Suffolk boars were told that these had to be "Thorough-bred" and that the pedigree papers had to be produced.

George Burrhus was superintendent of the swine department at the initial Walton town Fair held by the newly organized Delaware Valley Agricultural Society in 1887.

The class was for "pureblood Berkshire, Poland China, Chester White, Yorkshire, Essex, Suffolk or other pure-bloods." First place for a one-year-old or older boar or sow was still $3, and younger entries earned $2. The premium sow with at least four suckling pigs still got $3 and the second-place winner received $1. But by the next year the top premium had risen to $4 where it remained until about 1916 when it rose to $5. There it stayed through the 1940s. In 2000 a blue-ribbon sow with a litter earned $75, red, $50 and white, $35.

Until sometime after World War II every farm, in addition to a dairy, had a few chickens and pigs, but as dairies grew larger and the work more intensive, something had to give; on many farms it was either the pigs or the chickens, or both. At the Fair during the 50s and 60s, pigs were among the missing.

They returned in 1974 when Graham Jackson's sister, Ida Graby, became superintendent for a year. From that time until the present, there have been enough exhibitors to keep the swine department alive. Still, it is an area in which a few families have done a lot of the work.

Richard LaTourette and Ida Graby served until 1979, after which Ed and Jeri Brayman of Mormon Hollow Farm, Masonville, took over. They have supervised the department for 21 years. Richard LaTourette and his wife, Nancy, who are superintendents of the 4-H swine department, have served since 1977.

Much of the department's success depends upon the youngsters, says Jeri Brayman. Richard LaTourette points out that 4-H members show 90 percent of the swine, although many of these are shown in open classes as well. Both families have bred and shown swine for a long time, and their children, one after the other, have raised beef, sheep and pigs and exhibited them.

Eric Brayman in 1995 was a Triple-Crown winner, having achieved a championship for all three species. Eric's mother reports that the three boys, Duane, Neil and Eric literally "grew up" at the Fair. When he was in the Navy, Duane drove overnight from Chicago just to be there. The Brayman family's involvement in breeding and showing goes back at least four generations, and along the line, Everett Young,

Jeri Brayman's father, showed sheep and cows. Jeri started showing cattle when she was eight years old.

Both families, at one time or another, have had the novel experience of having a litter of piglets at the Fair, naturally drawing an appreciative crowd. Dick LaTourette says that for years he planned the breeding dates of his sows so that there would be a batch of new-borns for fair week.

Dick, who has also been a member of the fair board since 1982, began to participate in the Fair as a teenager. A member of the 4-H, he started by showing cattle and entering the tractor pulls. As they began to grow up, the LaTourette youngsters went to the Fair with swine, sheep and beef. The oldest son, Erwin, started his 4-H career in 1972, showing dairy cattle. He showed swine in 1974. In due course, Brian, the second son, was also winning blues for top-grade swine. Presently Nanette, Jason and Aaron joined the others in a variety of presentations.

Since 1979 to the present, the 4-H swine show has shown continued growth with only slight fluctuations, from 36 swine entries in 1979 to 74 in 2000 when 36 4-H'ers participated. In 1979, swine and sheep were included for the first time in the 4-H beef cattle auction, and this provided an additional incentive. The number of market animals being shown continues to grow.

At the beginning, swine were housed in temporary pens which had to be assembled beforehand and dismantled at the end of the Fair. An increasing number of entries made it desirable to build new pens, but after they were built, the new Paul Mattern Livestock Barn went into the construction schedule. What happened next, makes for an often-told, Paul Bunyan-type tale, but this one is true.

The new pens, with their supporting posts, had to be simultaneously lifted out of the ground and moved. About 50 strong people were rounded up; they did manage to lift everything, and they carried the whole unwieldy hodgepodge to the track where it all lay until the barn was finished. Then, using the same technique, everything was moved back in.

The new barn has provided a clean, modern, dry, and efficient show structure for swine, sheep, beef, goats, poultry and rabbits, all under one roof.

Photo courtesy of Jeri Brayman

Everybody wants to touch the cute black-and-white baby pigs. Ed and Jeri Brayman of Morman Hollow and their family were the proud exhibitors of this litter in 1992.

Photo by The Delaware County Times

A good brood sow can raise 12 to 14 pigs every litter.

Photo by Robert N. Renneck

A lick and a promise?

Photo by The Delaware County Times

Neil Brayman happily feeds his spotted pig.

Photo by The Delaware County Times

Board member Richard LaTourette chats with exhibitors at the 4-H swine show.

Photo by Robert N. Lennek

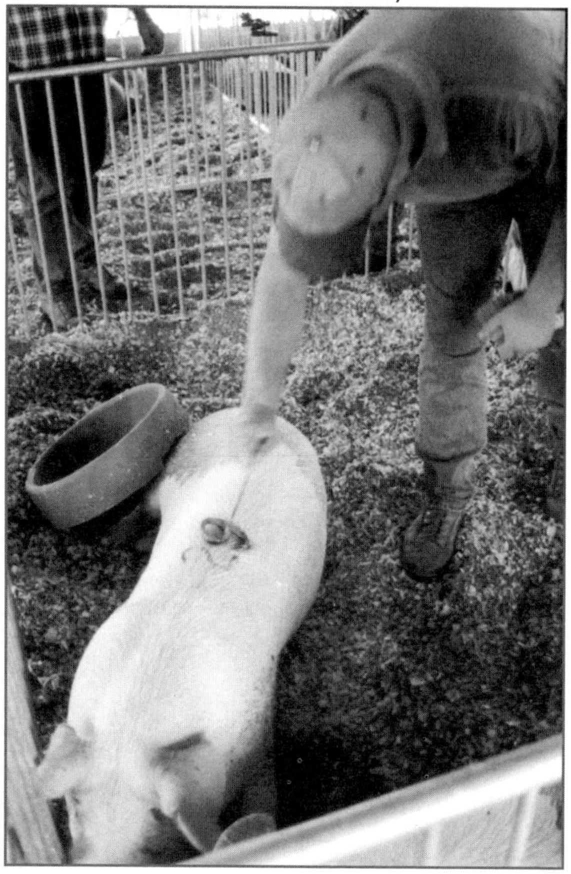

Good luck, Number 29! In the background is a section of the new metal pens that have made fair week so much more pleasant for both swine and people.

Photo by The Delaware County Times

Picture of a winning pig and a happy exhibitor.

Chickens Are Not What They Used To Be

Photo by The Delaware County Times

The Delaware County Fair has not had an open class poultry show since 1948. Some of the chickens that year apparently went home with an infectious disease and that was the end of open competition for premiums for a long, long time until a few people began to be interested in standard-bred poultry again and the 4-H began to offer programs which included exhibiting at the fair.

In recent years the number and variety of chickens at the Fair have slowly increased because the 4-H has a well-planned and extremely popular poultry program and because some of the hazards of the 40s have disappeared. At the same time, the once fervid excitement over gorgeous chickens has simmered down to a lukewarm interest on the part of a few still ardent breeders and keepers. Except for the commercial breeds, chickens are an endangered species!

Here's the story:

The earliest drawing of a chicken, that can be authentically identified as a chicken, is supposed to have been found on a Babylonian cylinder seal dated between 3,000 and 2,500 B.C.

In the New World, Columbus is said to have found wild chickens when he landed in the 15th century. And the Jamestown colonists and their successors brought domestic fowl with them.

Sometime around 1700 the French decided that poultry could be a source of profit all the year around by artificially hatching and rearing poultry.

Early in the 1800s, several large and hardy types of fowl—Langshans, Cochins and Brahmas were imported from China and the East Indies. The first poultry census in the United States was taken by the State Department in 1840, just a year before the first Delaware County Fair. According to the census calculation, poultry being raised in New York State at that time was valued at $1,253,413 (but the value of the dollar in those days is uncertain).[1]

The sparse records of the first Fair do not list chickens or ducks among the creatures on exhibit. However, this is not surprising since poultry raising probably had not reached the show-worthy stage in what was then a Catskill wilderness. Furthermore, proof of the growing interest in purebred fowl, the first American Poultry Show, did not take place until eight years later, November 15-16, 1849 at the Public Garden in Boston, Massachusetts. A total of 1,423 birds were shown there by 219 exhibitors; the attendance was estimated at 10,000, giving standard-bred poultry a highly successful start in life.[2]

A quest for beautiful birds would have been one reflection of the romantic turn of mind in those years. Nature and its manifestations were beginning to be seen, no longer as a threat, but as a wonder to be cherished. Along with this change in outlook came a very practical force, the basic competitive nature of human beings, which pitted the early poultry fanciers against each other. The result was an evolving specialty with a huge potential for growth. As better and better quality in poultry emerged, an unprecedented development of a whole industry was made possible, but as the industry matured, of course many fine species fell by the wayside. And that is where we are today. But, back to Delaware County.

By the time the Delaware Valley Agricultural Society was formed in 1887, the enthusiasm for standard-bred poultry was very high indeed, and Thomas Jamieson, who presided over Poultry, Class 5, had his hands full with the long list of entries. There were **Brahmas, Cochins, Dorkings, Dominiques, Gold** and **Silver Hamburgs, Black Spanish, Wyandottes, Langshans, White** and **Brown Leghorns, Houdans, Plymouth Rocks, Black Breasted Red** and **Duck Wing Games, Black, Silver** and **Golden Polish; Sebrights**, and a few **Bantams**.

While the first "American" breed, the Plymouth Rock, was shown as early as 1869, many others had been admitted to the American Standard only as recently as the 1870s and early 1880s. For instance, Andalusians, Hamburgs, Dorkings, Cochins, Leghorns, White-faced Black Spanish, Polish and light and dark Brahmas were admitted in 1874, the year in which the first Standard of Excellence was produced by a newly organized American Poultry Association. Langshans and Javas (highly popular chickens at the time) were admitted in 1883. The Silver Laced Wyandottes, which originated in

New York State, had also become a Standard Breed in 1883.[3] So, if nothing else, the excitement of having a new pursuit made chickens a big thing at the 1887 Fair and this excitement stayed at a peak for many years.

About this time, or a little earlier, some say during the late 1870s, a phenomenon called the "Stringmen" happened upon the county-fair scene. There is no doubt that they were to be found at the Fair in Walton, at least up to the early 1930s.

A "stringman" was, in a real sense, an entrepreneur. He was a person who either bought or raised a "string" of chickens and showed them at fairs all over the landscape for the purpose of winning the premiums. A complete string usually consisted of a cock, hen, cockerel and pullet in all of the large breeds and most of the bantams. Since the premiums were relatively small, it took an enormous number of chickens to profit, so an ambitious stringman might arrive at a fair with 400 to 700 birds. In the early days, teams of horses could be seen pulling wagons piled high with chicken crates on their way to a fairground. In due course, the big fellows used railroad boxcars and later trucks.

In discussions about them, stringmen are both honored and maligned. The stringman/breeder was more important to small poultry raisers than many people realized because he provided a source for selling surplus stock, as well as a means for comparing what was being bred and raised, thus upgrading the general quality of a breed. In addition, the stringman, because he raised so many types, was able to keep alive some of the less popular but interesting and useful breeds, some of which are now in danger of becoming extinct.

On the other hand, there were many complaints that the stringmen were careless and dirty, spread disease and took inferior stock to the shows. Furthermore, local poultry people grumbled that the stringmen, who came from far and wide—New Jersey, New England, Pennsylvania, even Ohio—were taking the premiums that should have stayed at home.

The best stringmen took great pride in their chickens and kept them clean and well fed and the coops attractive and brightly painted. But other coops were dirty and smelly with unchanged litter and neglected feed trays. With contaminated food and unwholesome conditions, there was every chance of spreading disease from fair to fair. It was the latter type of person that helped to spell the end of poultry shows, not only in Walton, but also at many fairs around the country.

A look at the Delaware County Fair of 1916 gives certain evidence of the stringmen's presence. The number of chickens was awesome. The premium book lists 14 sections for poultry, including chickens, turkeys, geese and ducks, with premiums at $1 for first place and 50 cents for second.

The 11 sections for chickens were divided into **American**, which included a huge number of Wyandottes and Plymouth Rocks, as well as Dominiques, Javas, Rhode Island Reds and Buckeyes; **Asiatic** with Cochins, Langshans and Brahmas; **Mediterranean** with 36 classes for Leghorns alone plus Minorcas, Anconas, Blue Andalusians, Spanish and Malays; and **Polish**, which had 32 classes. **Hamburgs**, with 24 classes, had a section of their own also. The section for **French** included Houdans, Salmon Faverolles, Crevecoeurs and La Fleche, and the **English** classes were for Dorkings, Orpingtons, Sussex, Cornish and Red Caps. **Miscellaneous** entries included Silkies, Sultans, Campines and Frizzles, and there were four classes for **Games**.

Bantams were becoming popular. There were two sections with Game Bantams entered in 26 classes and Ornamental Bantams in 80, making a total of 106 classes for bantam chickens alone.

The total number of classes for large fowl was 372, making 478 classes in all for large Standardbreds and Bantams in Walton in 1916. This is nowhere near the 9,300 entries at the State Fair poultry show in Syracuse in 1915, but it is a lot of chickens. It means that there must have been somewhere between 700 and 1,500 chickens at that Fair, hardly likely all of them local! Thomas Ford was the general manager and superintendent of this well-represented department.

Nearly 20 years later Vanley Blackman could testify to the presence of stringmen in Walton because he fed and watered their chickens. In 1933, when he was just a young kid, Van agreed to work in the Chicken Building. Between where the Grange Hall is now and the track, there was a small structure which housed the open class poultry. You went up two stone steps into this structure and there were wooden coops all along each side and the back, Blackman says. He had to sleep at the fairground during fair week and each morning he would water all the chickens and dole out the feed, which the fair association provided. After showing a Guernsey calf at the fair when he was no more than eight, he had developed an interest in poultry and was raising Silverlaced Wyandottes, a popular breed at the time. One of the stringmen bought a pair of these chickens from him, Blackman says. "They were in it as a business, and they knew what they were doing."

Vanley Blackman has, throughout his life, continued to be actively involved in the Fair. He was elected to the fair board in 1958 and was a member for nearly a quarter of a century. He served as its secretary in the 70s.

Sometime in the late 30s or early 40s the stringmen disappeared from Walton.

In a remarkable book entitled *The Stringman's Scrap Book*, published originally in 1974 and revised in 1979, poultry fancier Marion R. Nash recognizes the contributions that many stringmen have made toward the preservation of rare and beautiful poultry breeds. He also acknowledges shortcomings. Some stringmen had "no devotion whatever to the creation of high-quality birds," their only thought being to draw out prize money without considering any return to the system that fed them. But for others the joy in life was to breed the best that could be attained.

Stringmen came from all walks of life, Nash reports, and many "had the eye and the unfailing ability to pick out the good bird—the BEST bird!"

The book contains penetrating pieces by Maurice C. Wallace and Judge John Wunderlich of the American Poultry and American Bantam Associations. Wallace tells of a visit to the longtime and highly regarded breeder and string keeper, John C. Kriner. He looked at the "long, low, empty henhouse in Pennsylvania where Kriner and his birds had flourished in their day." He said: "Within me abode the sense of standing on hallowed ground." Kriner was the first breeder in this country to import Sebastopol geese.

Many pictures, some ancient and obviously rare, depict the stringmen and their flocks and their awards (mostly men, but a few women, too). The book contains newspaper clippings and lists of Fair wins as far back as 1901 and as recent as 1971, and, the highest tribute of all, a long biographical section describing the stringmen who bred poultry, bought and sold them, and traveled the fair circuit for a hundred years or more. Nash says: "They are very much a part of the folklore of these great United States...the stringman...is not only unique to America and Canada, but to the world, as he has not appeared anywhere else." Some were colorful characters. One of them allegedly traveled in a parlor car as he transported his "string" by train in the thirties from fair to fair. Most of his compatriots were in a boxcar with their poultry.

As time went on, boxcars went the way of the horse, and during the Depression many fairs reduced premiums or eliminated open classes, or both. Society was changing. For some years yet, backyard chicken flocks would be a common sight, and many small farmers would count on poultry for at least part of their income. But the era of the mechanized poultry business was coming, and as it came, raising poultry in a small way became unprofitable and to a degree unfashionable. The glamour of raising birds was fading. The people who would continue to raise standard-bred chickens would do so not for profit but because they loved them and took pleasure in improving the quality and beauty of their breeds.

New York had its share of stringmen. Included among those listed by Marion Nash are Billy Albers of Hicksville, Long Island, Helen Aylsworth, Brookfield; Harold Austin, Cambridge; H. A. Borgeldt of DeRuyter; Al and Gus Braun, Charles Clift, Perry and Rose Fish, Carl Hibbard and Harold J. Shuert, Syracuse; Ray C. Brown, Hampton; Charles Burmaster, Adams Center; Herbert G. Comstock, Pan Yan; Tommy Dorchester, Watertown; Harold Forward, Bouckville; Phil Hanson, Pulaski; John Hayes, Mount Upton.

Also, Howard Hidley, West Sand Lake; the Hoose brothers, Cherry Valley; Mrs. W. W. Howes, Old Chatham; Harry Light, Gilbertsville; Nettie Ellis, Cortland; Mrs. W. P. Masten, Pleasant Valley; George Montgomery, Cambridge; Robert G. Morris, Seneca Falls.

And, Brice Moore of Myown Farms, Grahamville; Bob Newbold, Holley, W.A. Smith of Otselic Farms, Whitney Point; O. A. Peacock and Roland Pierce, Canandaigua; Louis Perry, Clay; G. Archie Turner and his son, Robert, Horseheads; W. A. "Bill" Shafer, Oneonta; Ira Sincebaugh, Ithaca; Jesse Snow, West Stephentown; Arthur "Art" Schallenberg, Westernville; O. D. Spaulding, Burdett; Arthur and Bertha Traver, Troy; Earl Van Wormer III, Central Bridge and Lee Williams, Adams.

A few stringmen are still active. One of these is a neighbor in Mount Upton, John Hayes, whose string of chickens, ducks and geese is, at this writing, still winning championships at county fairs, at the annual Eastern New York State Poultry Association show in Cobleskill, at the Yankee Fall Classic in Syracuse, and at the New York State Fair. So they aren't all gone, but their ranks have been dwindling. John Hayes started showing when he was ten, served in World War II, moved to Mount Upton in 1954, and has been showing chickens, ducks and geese ever since. Noah, he once said, may have been a more important Biblical personality than we are apt to admit.

Through the 20s and 30s many local breeders competed in the open poultry classes at the Delaware County Fair and they were consistently joined by 4-H members. The 4-H started exhibiting poultry in 1922, the year John Lennox arrived as the county 4-H agent.

When he came, in the early part of that year, the 4-H was only four years old. Exhibitions of poultry and livestock that summer at both the County Fair in Delhi and at the then town fair in Walton was one of the first things Lennox engineered in a tenure which lasted for 21 years. Poultry-culling demonstrations and a parade were highlights of 4-H participation that first year.

By 1925 there were twice as many 4-H poultry members as there were dairy members, and all through the 20s poultry was high on the list. In 1931, 220 birds were shown at the Fair. By 1934 some 171 poultry flocks were registered with the 4-H.

In his 1986 "History of the Delaware Valley Agricultural Society," Richard Wakeman tells the story of an old-time poultry exhibitor who would prime his hens and roosters. He'd buy a pound of bologna when he took his milk to town and feed the chickens a piece every time he approached them. His rationale was that the judge liked to have the birds look at him while he was judging. The strategy must have worked, because these chickens are said to have earned many a blue ribbon!

Although World War II brought a host of distractions, the 4-H seems to have maintained its interest in poultry. A group of youngsters trooped down to the 1943 New York Poultry Show at Manhattan Center, New York City. There they were allowed, for the first time, to exhibit "production poultry"—Rocks, Leghorns and Rhode Island Reds. The same year Wallace Rich provided White Leghorns at his poultry farm in Hobart for a 4-H poultry judging contest. A newspaper clipping after the 1943 Fair noted that "a small, but select 4-H Club poultry show attracted the attention of visitors." David Edwards of Delhi and Roy Mains of Sidney Center showed the champion cock and hen, respectively. The open classes, however, were showing fewer and fewer entries each year. The 4-H was beginning to be the lifeline on which poultry depended. But the emphasis was on birds for production—eggs or meat. It would be a while before they showed anything more than Rocks, Reds, Hampshires and Leghorns.

In 1944 the follow-up report was that "Exhibitors in the 4-H poultry show encountered considerably more competition than a year ago." David Edwards and Roy Mains again took the championships.

A new column entitled "Poultry Notes" by Professor R. C. Ogle of the poultry husbandry department, Cornell University, began to appear in the 4-H Newsletter in 1945. He also arranged with the Cooperative GLF for a state 4-H Club home egg-laying contest. It was designed to encourage 4-H'ers to keep laying flocks throughout the year in order to show that keeping accurate records would result in greater profits. In 1946, Professor Ogle wrote a column on pullorum disease, an infection passed to chicks from the eggs which hatched them. In the premium book for that year's Fair, a new kind of notation appeared: "No birds with diseases, lice or mites will be allowed."

In 1947, there was another trip to the Madison Square Garden Poultry Show, billed as "one of the largest poultry shows ever held in Madison Square Garden." The sponsor, Crawford Brothers Feed Company, sent David Edner of Delancey and Vincent Trimbell and Glen Kaufman, both of Bovina, who had shown County Fair champion fowl.

If New York City in 1947 had "one of the largest shows ever," here in Delaware County the reverse was true. Interest in poultry had declined to the point where at the Fair in 1948 there was only one open class exhibitor of chickens and two of ducks and geese. "I guess we weren't much for fancy birds around here," commented one observer who remembers that year. The active participants were the 4-H members. After the Fair, however, several 4-H exhibitors' parents suffered severe losses from disease in their poultry flocks, infected, it is thought, by birds that had returned home from the Fair.

Exactly what the disease was remains something of a mystery. Some suggest that it might have been pullorum disease which Professor Ogle had mentioned in his column; others feel that it was probably Newcastle Disease about which there had been widespread warnings. It was a respiratory infection similar to bronchitis which produced a "peculiar cough and gurgle and the dropping of eggs on the floor" instead of in nests. It had almost never been encountered in Delaware County before the Fall of 1948, according to an article in the Delaware County *Farm and Home Bureau News.* There was a vaccine, but "no satisfactory remedy for a flock that once has the disease."

In any event, before the Fair of 1949 this notice appeared in the premium book:

Because of the danger of spread of disease among commercial poultry flocks of the county, it has been decided to discontinue the open class poultry exhibits.

In order to prevent a reoccurrence, (of disease among flocks the previous year) *money previously used for open class poultry has been allotted to 4-H Club poultry to be used to recompense 4-H exhibitors for birds shown under a new plan whereby all poultry will be sold for meat purposes at the end of the Fair.*

The 4-H Club *News* told its young readers that they "would be paid the regular market price per pound for the birds plus $1 for each pullet or hen to compensate for loss of egg production." The next paragraph read: *We are sure that all exhibitors will welcome the change, because it not only protects their flocks, but also makes it unnecessary for them to call for the birds at the end of the fair.*

Years passed during which, many times, no fowl of any kind were shown.

Then, sometime in the 80s a glimmer of interest returned. Beth Adams of Hobart, whose father had raised chickens and exhibited them, appeared on the scene and her enthusiasm for beautiful chickens was communicated to her daughter, Laura, and her friends. The result was that within a few years there was once again a highly varied and interesting section of standard-bred poultry. Beth Adams has retired as superintendent of this department and Dan Grant of Hobart has taken over. As the 2001 Fair comes into view, the 4-H will have a commendable poultry display.

There is a gnawing problem, however. Raising chickens is no longer profitable for the small operator and raising them just for the joy of it no longer seems either glamorous or fashionable.

Fortunately, the situation has not gone unheeded. In 1977 a group known as the American Livestock Breeds Conservancy arose which works to protect more than 100 breeds of cattle, goats, horses, asses, sheep, swine—and poultry—from extinction. Based in Pittsboro, North Carolina, it conducts a periodic census of poultry breeds, issues alerts about endangered species and encourages the breeding and raising of those species.

A second organization, the Society for the Preservation of Poultry Antiquities, publishes a yearbook listing poultry by breed together with a list of breeders throughout the country who are registered with the Society, thus offering an outlet for purchase, exchange, questions and information.

But as John Hayes says, "You can't save chickens by telling people they have to. It has to come from appreciation, admiration and desire." For the Delaware County Fair once more to show rows and rows of beautiful chickens in future years, a lot of things have got to change.

Cornell Cooperative Extension Archives

During the 1950s, Sears provided baby chicks for 4-H projects. This is John Adams of Hobart with his white Leghorns which were one of the most popular breeds.

Photo by Robert N. Lennek

The incubator with the transparent top, through which hatching chicks can be viewed, fascinates all who watch and proves that eggs do not just come from the store.

The Fascination of Newly Hatched Chicks

Photo by Robert N. Rennek

. . . soft and fuzzy and warms the heart as well as the hand.

Photo by The Delaware County Times

Beth Adams, who was superintendent of the poultry department for several years, fluffs up a Cochin. The breed came from China about 1845 and left the public flabbergasted by the chickens' size and appearance. They have also been referred to as "Chinese Shanghai Fowl."

Photo by Robert N. Lennek

All kinds of ornamental chicks can be found in the poultry department. The little tyke is a Light Brahma bantam.

Photo by The Delaware County Times

Three winners of blue ribbons in the bantam and large-fowl categories smile for the camera.

Photo by The Delaware County Times

The 4-H has been an important contributor to an ongoing poultry department. The Grand Champion Large Fowl in this picture, taken sometime during the early 90s, is a white Australorp held by Laura Adams-Durgin, with the Reserve Grand Champion a black Hamburg exhibited by Silas Carey.

Photo by The Delaware County Times

A blue-ribbon Silver Laced Wyandotte hen looks perfectly content, nestled in the arms of her bright-eyed owner. The Silver Laced, parent variety of the Wyandotte family, originated in New York State.

Photo by The Delaware County Times

A Buff Laced Polish cockerel. Polish fowl, with their elegant, feathery heads, are mentioned as a purebred as early as the 16th century, and they are considered one of the most ornamental and extraordinary types of chickens.

Photo by The Delaware County Times

Portrait of a triumphant exhibitor and her prize-winning rooster.

Photo by The Delaware County Times

Now, tell me something about how you raised your chickens. . .

Although that probably was not their conscious motivation, the "stringmen" of the early 20th century were instrumental, to a degree, in preserving many breeds of poultry that have since tended toward extinction. Stringmen, who bought or raised "strings" of chickens and traveled wide distances to show, sell and trade them, could arrive at a country fair with 400 to 700 birds. Evidence of this comes from a volume entitled *The Stringman's Scrapbook*. Here, a large truck is beginning to discharge its poultry entries at a fair, possibly sometime in the 1930s.

This remarkable and historic picture, taken about 1910, shows one of the more famous stringmen, Art Quade, taking his string to a county fair by horse and wagon.

Rides, Rides, Rides

Photo by The Delaware County Times

The year was 1934 and it was the depths of the Depression. The fair board was looking for something new to brighten things up and in the course of exploring the possibilities, board member C. O. Field turned up a remarkable show business family: The Mighty Reithoffer Shows.

For perhaps the first ten years or so of its existence, the Fair had been synonymous with agricultural exhibits. But by the early 1900s, there began to be other attractions. The side shows came, and gambling booths and luck-and-skill games and rides. Popular as these features were, they did not always have a reputation for honesty and good clean fun. To find a carnival that would fit in with the standards of the Walton Fair was a challenge, and Walton in 1934 was lucky.

The Reithoffer family's history in America began in 1898 when the patriarch, Julius, bought a steam-driven Merry-Go-Round and started to give itinerant shows. The Merry-Go-Round first was moved by horse and wagon from show to show, but eventually, like the circus, it hit the rails.

About the time that the Mighty Reithoffer Shows were at the peak of success, Julius' wife, who was only thirty-five, fell ill and died. Julius became so despondent that the business would have fallen apart had not his sons, Pat, sixteen; Julius, Jr., fourteen, and Eddie, twelve, struggled to keep everything together. And they succeeded. In doing so, they hit upon a new idea: Move into areas where trains did not go. They bought a fleet of new Packard trucks with rubber tires, and soon they had 50 of them.

In 1929, after the stock market crashed, Pat watched various carnivals go broke, and he made another strategic decision: Break the show into sections. Thus, by 1934 when the show came to Walton, Reithoffer was in sound shape and could look forward to more than half a century of sending the various units on the road in every direction.

After World War II, Pat's son, P. E. Reithoffer, Jr., a pilot in the Air Force, returned home and started to reorganize the business. After purchasing new equipment and surplus trailers and trucks from the Army, he expanded the route until the company had fanned out to ten small shows, thus allowing expansion but diversifying the financial risk. Larger and larger fairs were booked until Reithoffer was to be seen at well over 30 locations during the season and had gained the distinction of being the oldest known carnival in the country. In the year 2000 Reithoffer had three shows on the road to play approximately 90 dates a year. Pat Reithoffer III and his wife, Betty, head the organization at the present time. From summer headquarters in New Albany, Pennsylvania, they make the circuit, visiting their shows.

After a while, the routes were split up again, and Pat Jr.'s daughter, Arlene Reithoffer Goodman took over the Walton scene with Brooke Evans as business manager. Arlene Goodman died in 1996, leaving Evans to oversee Walton. He visited this Fair for the first time somewhere in the 60s and decided that "it was one of the prettiest sites you could hold a fair on." He started work in Walton in 1970, having been released from the Navy. He had served in Vietnam and earned a Bronze Star.

Brook E. Evans has been part of carnival life since he was fifteen. His home is in Kingston, Pennsylvania, with his wife, Lynn, son T.J., who is just finishing second grade, and daughter, Haille, in Kindergarten. But he says his real home is on the road; the whole family travels in a trailer from March to October. The couple met while both were working for the Reithoffer organization.

Mr. and Mrs. Pat Reithoffer, Jr.

He says: "When I first started coming here there was nothing but tents. Now there are solid buildings for the cattle and horses and for many of the commercial exhibits, which seem to be more numerous each year.

"Our Midway has grown along with the rest of the Fair, and the crowds have kept right up with the growth. The only thing that has stayed the same is the atmosphere. Delaware County has a *country* fair. They are very proud of that fact, and they work hard to keep it that way."

Every Thursday for the past decade Reithoffer's Rides has been host to groups of mentally and physically handicapped children who might otherwise not have a chance at the rides and games.

As a carnival, the Reithoffer show consists of rides, games, food and occasionally side shows. The old type of carnival side shows, the ones that caused such consternation in the middle 1900s, are no part of the present scene, however. Barkers do not call attention to provocative dens. The "gamblers and half-naked human sharks" are missing. Fat ladies and freaks are no more, a more compassionate society having decided that it was cruel to display unusual people and gawk at them. Instead there are the ring-toss and similar skill games where one can win a huge teddy bear or stuffed panda. But you can't win a bag of goldfish any longer.

The rides are apt to change slightly from year to year, with a few new ones added to the old familiars. Some of the famous rides are the Skywheel, which is a new name for the old ferris wheel; the Music Fest, which is a fast round-and-round thing; and the Hi-Roller, which is a ride that goes up and down, actually a large train that runs around a big rectangular track and is upside down part of the time, and then the track itself spins. People—not just kids—love them all.

Photo by Robert N. Rennek

Reithoffer's Rides, Tent City, the huge cattle barns, and the new Grandstand.

Walton Historical Society Photo

A little girl, in 1941, could ride round and round in a spiffy car like this.

Walton Historical Society Photo

The Ferris Wheel is still an old stand-by, but we are now wearing pantsuits.

Photo by The Delaware County Times

A put-put boat or a shot at one of the dart games makes the Midway a favorite place to spend time at the Fair.

Photo of son, Dick, by Robert N. Lennek

So what can I get into next?

Photo by Robert N. Lennek

The merry-go-round is a source of perennial pleasure.

Photo by Robert N. Lennek

Whe-e-e-e....

Photo by Robert N. Lennek

The Whisper Cart promises interesting adventures.

Walton Historical Society Photo

The 1948 Midway featured towering rides, as well as the old-fashioned Merry-Go-Round. Notice the skirts on the girls. . .

Smash 'em Up: The Demolition Derby

Photo by The Delaware County Times

The Demolition Derby may have been a child of the 60s, but the yen for thrills that gave it life goes way back to the days just after the Depression when this nation's love affair with automobiles was starting to blossom and daredevil young fellows were bent on outdoing each other by performing hair-raising stunts with cars.

As early as 1940 one of the more famous of these thrill drivers arrived, Lucky Teeter, whose name still brings a smile of reminiscence to those who witnessed his daredevil performances.

The year before the War ended, B. Ward Beam's International Congress of Daredevils came to Walton and offered Happy Hapgood, who did "what is considered the most dangerous stunt in the thrill game." He drove his motorcycle at top speed through a 20-foot tunnel of flaming excelsior. It was one of a dozen "great thrillers" offered on Thrill Day at the Fair, including a black sedan that jumped over a bus.

The next year the spotlight was on the "world's champion Hell Drivers," offering "death-defying auto crashes," and "nerve-shaking crash roll-overs."

Catering to the deep-seated desire of human beings to tear things apart, the smash 'em up thrill shows were a guaranteed sensation, and the cry was for more. So early in the 1950s came Joie Chitwood and his world champion daredevils who "smashed and crashed $150,000 worth of new Ford Sedans, leaping cars through walls of death"! He was probably the most popular of the daredevils in those years. They included Jack Kochman and his

Hell Drivers, who billed himself the successor to "the late, great Lucky Teeter," and Buddy Wagner and his Cavalcade of Auto Daredevil Champions, and Joie Chitwood, whose name is still a familiar one at the Fair. Joie Chitwood first presented his Thrill Show in 1966.

When delegates from the fair board went to the New York State Association of Agricultural Fairs' annual trade show held in conjunction with its yearly meeting in 1968, they learned about something new. Up from the south had come the concept of a whole fleet of cars driven by local fellows bent on crashing and smashing each other. It sounded good. It was called a Demolition Derby.

Robert Butcher of Downsville agreed to be chairman of this new thriller, and Clayton Stanton of Walton became vice chairman. Both men were race car enthusiasts; Bob Butcher ran a garage and he was also a stock car racer. Later on, he was very much involved in the Five Mile Point Speedway near Binghamton.

Amos Finch and Dusty Doyle were the announcers for the two derbys held on Monday and Wednesday nights in 1968.

Looking back at the early years Amos Finch recalls: "I remember working with Bob Butcher when the Demolition Derbys started. I did the track announcing. I remember it always seemed as if we were left standing on the old stage covered from head to toe by mud when the event concluded each year. Then Gary Tweedie (God bless him) came along to do the announcing and I got to sit in the stands and enjoy the event. The first couple of years we did broadcasts from a tent and then we switched to Mobile Homes as our studio on the fairgrounds." Amos Finch and Myra Youmans, who had sold their interest in the Delaware County Broadcasting Group and were retiring, were honored by the fair board at the 2001 annual meeting for 45 years of promoting the Fair. Finch told the board that his association with the Fair had begun in the late 40s as a member of the 4-H. "My great-grandfather, Fowler Finch, moved to Walton after he retired and during the summers in the late 1920s, he worked as a ticket taker at the main gate of the Fair. My first broadcast experience at the Fair began in 1955 working with Dick Hoyt for WDLA."

In 1968 each heat winner got \$50 and the final winner, \$100. The most popular driver in each heat won \$25, and for the final heat, \$50. Today the winner of each heat wins \$150, the second-place winner gets \$125, and the grand prize is \$350. Second place receives \$250, and all winners receive trophies. The first-place winner, who gets to compete at the State Fair, can bring home a top premium of \$1,500 so there is a terrific incentive to drive a Derby car.

Sometime in the late 70s the Women's Movement arrived at the Derby. With the girls clamoring to compete, Derby officials activated a Powder Puff heat, which led to a remark by one fair observer that "we sure have some new kinds of girls here!" They pitched in with such vigor that soon the Powder Puff designation was tossed out, and guys and girls now compete equally. Has there been a female Grand Champion? Two of them! Amy Moody of Walton and Michelle Palmer of Hamden both made it to the top and went on to the State Fair.

After Bob Butcher died in November, 1971, the Fair asked his wife, Joyce, (who is now Mrs. Richard Webler) if she would help. Now, nearly 30 years later, she is still helping. As vice chairman, she receives all of the applications, sets up the events, and she can be seen onstage as the prizes are awarded. Both of her sons, Robert and Mark, and her daughter, Stacey, have taken part in the Derby. Mark is now part of the Derby team, in charge of scheduling and setting up the program.

Following the death of Bob Butcher, Stan Budine took over the Derby, and he ran it until his own untimely death in April, 1981. Gary Tweedie has coordinated the event since then and acted as announcer.

At its inception, the Derby raised some controversy because of its potential danger. A major concern was for the safety of the drivers so a list of tight rules and regulations was drawn up. Over the years, the Fire Department has stood guard over the rumpus with its equipment ready to go. Eligible cars are any American hardtops or station wagons. All glass and chrome must be removed from these cars, gas tanks are moved into the trunk and the tank can contain only five gallons of gas.

Drivers, who supply their own cars, must be at least eighteen years of age and must wear helmets and seatbelts. Anyone found using alcohol will be disqualified. Cars are barred from the track until their heat is called; only drivers and mechanics are permitted in the pit or on the track.

The Demolition Derby is open to residents of Delaware County only. Winners who go on to the State Fair in Syracuse are representatives of this county and the words "Delaware County" are painted on the side of their cars. The State Fair Derby is apt to attract pros as well as amateurs so winning the top spot is an enormous challenge.

Writing in the *Delaware County Times* during the summer of 1999, Peter Mauer of Bloomville spoke for multitudes of fairgoers when he said: "We love going to the Demolition Derby. It is such fun to watch all the cars smash into each other."

The Derby, now in its fourth decade, is still one of the biggest crowd-pleasers the Fair offers to its

Walton Historical Society Photo

The Joie Chitwood Thrill Show has been around for a long, long time. The early thrill shows of the 40s and 50s whetted the public's appetite for something like the Demolition Derby, which was acclaimed when it arrived in 1968 and has been a cornerstone of the Fair ever since.

Photo by The Delaware County Times

Today it pays to be in the thrill show yourself. Prizes for Derby winners can go as high as $350.

Photo by The Delaware County Times

A small conference, but everything seems to be okay, thank you.

Photo courtesy of Irma and Leo Pierce

Before the destruction, contestants go to all kinds of trouble to make their cars look special. Mark Pierce created this one.

Photo by The Delaware County Times

Gary Tweedie has been the announcer and chairman of the Demolition Derby since 1981, following Bob Butcher and Stan Budine. Here Lindy Klinger is presenting an award.

Photo by The Delaware County Times

Pick up time.

Photo courtesy of Lorinda Pierce Backus

Perhaps it's time to go home. . .

Photo courtesy of Irma and Leo Pierce

After the Derby, many a back pasture has its graveyard of skeletons.

Home and Garden, Arts and Crafts

Photo by The Delaware County Times

Once upon a time there was a Ladies' Department in a Ladies' Building. In it were exhibited all of the projects ladies did at home, and these were divided into four categories: Fine arts, domestic manufactures, fancy articles and culinary. These headings still characterize the many useful and artistic items shown at the Fair, but the list has grown and changed, and the categories are no longer just for "ladies."

Domestic manufactures, fancy articles and culinary have become the department called Homemaking, which, in 2000, offered a total of 445 opportunities to enter all kinds of products from the sewing room, kitchen or garden.

There were 186 classes for various kinds of needlework and crafts, another 168 classes for baked goods, including pies; 18 classes for candy and maple syrup products, and 73 for canned goods, jellies and jams.

There was also a separate department called "Flowers," offering 108 classes for cut specimens, arrangements and potted plants, and a section for children. All of this was under the chairmanship of Fran Robinson and shown in the Grange Building.

In addition, Bunny Houck and Karen Kemp were superintendents of the department called Table Settings which was comprised of 16 classes open to both men and women.

The Art and Craft Show, chaired by Kathleen Moore and Arlene Lurenz, held in the 4-H Building, had taken on a life of its own long before 2000, and it has become a mecca for all kinds of artists from a wide area.

In contrast, at the first Delaware Valley Agricultural Society Fair in 1887, there were 138 classes entered by the ladies in the four categories of fine arts, domestic manufactures, fancy articles and culinary. Truly, the Fair has grown.

The articles entered in those days would have included "domestic manufactures" such as a piece of rag carpet, wool stocking yarn, a patched garment or a "plain table mat." A fancy article might have been a hand-embroidered nightdress or pillow sham, an embroidered piano spread, drawn work, paper flowers or a tea cozy.

Today the list is longer and includes all kinds of embroidery, crocheting, tatting, cross-stitch, crewel, needlepoint, bargello, knitting, macrame and weaving. There is a section for rugs, quilts, latch hook, sewing of all kinds–and Christmas ornaments.

Baking? As might be expected, there are many more kinds of breads, cookies, cakes and pies on the list.

In 1887, there were only four categories for flowers: Floral design or ornament, collection of cut flowers, a bouquet, and the "largest display of flowers by one person." Today, you name the flower, and it probably can be found on the premium list.

As for arts and crafts, it is not clear whether the Art Show can claim the initial Fair in 1887 as its genesis, or whether it began a year later, but there is a record to show that in 1888 the work entered was collectively entitled a "Fine Arts Show."

The scheme of things for items in the Ladies' Building remained fairly constant through the end of the century with, usually, a male leader. The Rev. Scott B. Rathbun and the Rev. G. D. Ramsay, Mrs. John Olmstead and Mrs. E. L. Guild were the superintendents in 1887. The Rev. S. G. Shaw and Miss Fanny Howland took over in 1888. William A. Shepard became general superintendent in 1890, and the Rev. S. G. Shaw returned, assisted by a large committee, in 1891.

By 1909, however, this group of exhibits had become "Department K–Domestic," with Mrs. John M. Eels, general superintendent and manager, and Mrs. Ray D. Churchill, secretary.

By 1914, Mrs. W. C. Rogers had become the superintendent, with Miss Janie Launt, secretary. This was an era when hemstitching, drawn work, lace work, tatting, knitting and silk embroidery were popular pastimes, and home sewing was at a peak. The list of entries seemed endless. Also, the classes were divided into "amateur" and "professional." The list of baked and canned goods, pickles and jellies was equally long. There were also five classes for wine–blackberry, elderberry, raspberry, current and grape, but oddly enough, no dandelion. Plants and flowers had earned a department of their own in 1914: Department I, chaired by Mrs. George E. Robinson.

Since the premium books for the 1920s cannot be found, it is difficult to tell when this department changed its name, but by 1934 it had become Homemaking, and that is what it has remained.

Bertha Lambrecht, whose name appears in the premium books over a period of years, was the Homemaking superintendent in 1934 when she also headed the Art Department. Lydia H. Wheat was head of Department J—Flowers. By the middle 40s, Kathryn Moore and Mrs. Sheldon Hoyt were overseeing Homemaking, with Mrs. Tyler Sherwood in charge of Flowers and Dorothy Hawes, Art.

In the middle to late 40s, there was a section entitled "remodeled garments," in which sewers could win prizes for making a woman's suit from a man's, a child's snowsuit from a coat, a sunsuit from a man's shirt, or a hat and scarf set from a worn garment. Even though World War II was ending, the Depression was not quite over, and this was farming country where a little had to go a long way.

In 1955, the canned goods section seemed to have been especially long and there were three classes for "Meat." But no wine.

The 1957 Fair, which was held during the reorganization, was a big one for the Homemaking Department, especially needlework and handwork: Dining tablecloths, five-piece luncheon sets, bridge cloths and napkins, dresser scarfs, centerpieces, sheets, pillow cases, guest towels, bedspreads, quilts, afghans, clothing, handkerchiefs, rugs, knitting, crocheting, Swedish weaving, basketry and reed work, jewelry and embroidery. There was a small section of six classes for art. For some years art entries were included in the Homemaking department, and Art, as a department, disappeared. Mrs. William Tweedie of Mundale was the Homemaking superintendent, and she and her husband were in charge of Flowers. They worked together in this way until 1974 and were followed by people such as Mrs. Glenn Bowker in 1975, and Robbie Jean Rice in 1979.

By 1965 the Art categories under Homemaking had grown from one section to four with an added heading for ceramics, including sculpture, metals, stains and decals.

By 1974 the Art Show had officially arrived, and that is another part of the story. All this time, too, the flowers, as Department J, bloomed on the long tables in the Grange Hall in an ever-increasing profusion.

In the color-coded premium books of the 1980s, Department H—Homemaking, begins to look much as it will in 2001. The superintendent from 1981 through 1996 was Candice Russell of West Brook. Mrs. Russell, working with Diane Dreibelbis, set up computer records of the homemaking division. Record keeping for exhibitors of this department had previously been a monumental task because of the number and complexity of the entries. Noticing their work, the fair board in 1991, entered the computer age itself by computerizing all of the Fair records.

During all this time, what has come to be a well-regarded, county-wide Art Show was slowly being built.

In the reports of fairs held by the Delaware County Agricultural Society in various parts of the county between 1841 and 1887, there is usually a reference to art work. In 1869, for example, when the fair was held in Walton, there were oil paintings, "pastile painting," and drawings in crayon, pencil or pen.

At the first Delaware Valley Agricultural Society fair in 1887, the entries could include oil paintings, crayon, pencil or pen and ink sketches, watercolors, handpainted china, architectural drawings, and photography. There was also a class entitled "combined exhibit of paintings of all styles."

The "Fine Arts Show" of 1888 ran for only one day, but then the whole Fair had only three days of actual activity. Mrs. B. G. North, the arts chairman, and her helpers, hung romantic landscapes, winter scenes, marine views, animal heads, portraits, and flower paintings in oil; drawings in crayon and pen and ink, watercolors and pencil sketches. An exhibitor also could enter architectural drawings, a hand-painted lamp screen, a painted banner, mirror or mirror frame; fire screen or fire board, a toilet seat, blotting-book, fan or photograph holder. The premiums ranged between 50 cents and $1 for first place, and from 25 to 50 cents for second, the same as the previous year. In both years, Fine Arts was the first category listed in the Ladies' Department.

In 1895 a new category was added, one for professionals and teachers. (In the middle 60s, when art was part of the Homemaking department, the rules specified that "Professionals and teachers are not eligible to enter in any classes..." But when Art became independent again, this barrier fell.)

The categories were expanded in 1900 and divided into two divisions: "Life and Nature" and "Copies," but two years later, the show suddenly had grown smaller. Its records for the next 20 years, like those for the Fair in general, are missing.

By the 1930s, Art had become a Department. Its superintendent was Bertha Lambrecht and it offered sections for both professionals or teachers and amateurs. Entry, however, was open to residents of Delaware County only. With Mrs. Samuel Pond the 1937 superintendent, a division was added for "original architecture and landscaping." The fair board was working on a building project and the art department accepted, as exhibits, plans for Fair buildings and for development of the Midway. Also submitted were plans for houses, barns, garages, cabins and commercial buildings. A pottery division was formed and photography was expanded to include scenes and activities in Delaware County, buildings, livestock, and pictures of the Fair.

A Students' Poster Contest also had its beginning in 1937, open to pupils from fifth grade through high school. First prize was $3.

Mrs. C. O. Field headed the Art Department in 1946 when it added a Junior Division for students in grades 1 through 12. Any plate or project created during the school year, including mechanical drawing, could be submitted. This was expanded in 1952 to include painting, drawing, design, sculpture, and prints.

Beginning in 1970, an array of fairly well known artists began to be interested in showing at the Fair and selling their work. The shows, held in a tent for two days, were a great attraction, but after a few years, the group dispersed, and the show again became small. In 1974, with Charles Locherer, the wood carver, Mrs. Locherer and Robert Lynch at the helm, the art show consisted of three classes: "oil paintings and water colors, other media, and crafts—basketry, ceramics, sculpture, etc." During these years it was billed as "the annual Delaware County Fair Art Show" and it was held for two days, open to amateur and professional artists in Delaware and adjoining counties. Premiums were $7, $5 and $3, with $12.50 for the first "most popular" and $7.50 for the second.

The Delaware County Council of the Arts began to be interested in the art show in 1980, and in 1981 Mrs. Marcia Kozubeck, recognizing the need for a forum for local artists, proceeded to revive this event and turn it into a county showpiece. She talked the fair board into allotting space in the Cooperative Extension building, where it remains today. A premium book cover contest, which has attracted entries from school children throughout the county, was begun in 1982 under the leadership of Mrs. Kozubeck. Students in grades Kindergarten through 12 may compete. Three grand prizes, worth $16.50, are chosen from first-prize winners, ($11); they appear on the front and back covers and on the year's poster. Other prize amounts are $5.50 for second, $3.30 for third, and $2.20 for honorable mention.

For several years in the middle 80s, a one-day, outdoor, juried arts and crafts sale called "The Catskill Corner," was held at Fairway Street and Tweedies Boulevard, around the corner from the 4-H dorm building.

Recent shows have highlighted demonstrations by various artists who shared techniques and encouraged beginners. The most recent development is the inclusion of computer art. Today there are sections for professional and amateur artists and juniors, with the top premium for professionals $27.50 and for amateurs, $12.10. A far cry from the $1 prize of the early years.

Artists of note whose works have been shown at the Fair include: Burt Townsend, who has painted many historic sites, notably the Delhi County Courthouse Square for the 175th anniversary of that village; Anatoly Shiapak, a native of the Ukraine, whose works have been shown at international art expositions and in New York City galleries; wildlife artist Ward Herrmann of Delhi, and Charles Locherer, sculptor and wood carver of Walton, whose works are to be found in many area homes.

Kathleen Moore and Arlene Lurenz have chaired the arts and crafts show since 1987 and are assisted by a large committee. They, and the Fair, are in a strategic spot as the world of music, fine and practical arts, writing, and drama increase in interest and visibility in central New York State.

Stephanie Ehrhand
Grand Prize – Grade 9
Delaware Academy

Photo by The Delaware County Times

Potatoes, tomatoes, onions, cauliflower, squash, broccoli and many other carefully raised garden products can be seen at the Homemaking Department's exhibits in the Grange Hall.

Photo by The Delaware County Times

Table Settings have a long history of popularity at the Fair. "Welcome Home, Soldier" is the title of this one.

Photo by The Delaware County Times

Florence Tyler of Hamden has won accolades far and wide for her quilting artistry, and she has been a county Fair judge for a long time. Here she contemplates vests made of men's ties.

Fair Book Cover Contest Winners 1986 & 1992

In the Centennial year, 1986, the grand prize winner was Ryan Hughson, Townsend School, Walton.

Amy Joy Ziemba, in Grade 4 at the Downsville Central School was the winner in 1992.

Fair Book Cover Contest Winners 1994 & 1998

Grand prize winner in 1998 was Duane Dubois, Grade 9, Hancock Central School.

The winner of the 1994 contest was Megan Andrews, Grade 2, Charlotte Valley Central School.

Fair Book Cover Contest Winners 1999 & 2001

Adrienne Sangastiano, of Grade 9, Walton Central School, was the grand prize winner in 1999.

Katie Glasner, a 9th Grader from Townsend School was the grand prize winner for 2001.

Appendix

The Delaware Valley Agricultural Society Officers and Directors 1887 - 2001

This is the first list ever to be compiled of the fair board officers and directors. The list covers the period from the inception of the Society to 2001. However, a few of the Premium Books, from which the names were gathered, appear to have been lost.

For the first 35 years or so, the main roster was composed of Walton residents, augmented by a varying number of vice presidents from surrounding areas in Delaware County. In all cases, members are from Walton unless otherwise noted.

1887
President: C. B. Bassett
Vice Presidents: Joseph Harby, Jr., William C. Doig, William C. Rogers
Secretary: H. S. White
Treasurer: J. H. Townsend
Directors: Robert Harby, E. R. Howland, Alexander Tweedie

1888
President, C. B. Bassett
Vice President: Joseph Harby, Jr.
Secretary: H. S. White
Treasurer: J. H. Townsend
Directors: William C. Doig, William C. Rogers, Alexander Tweedie, T. H. Thomson, the Rev. Scott B. Rathbun, R. Garrison
Additional out-of-town vice presidents were: Capt. W. Hymers, DeLancey; S. S. Gregory, Cannonsville; William E. Combs, Hamden; Beliver Radeker, Colchester; Abram L. Boice, Trout Creek; Walter Miller, East Branch; William Tweedie, North Hamden; Hon. Irving Holcomb, Trout Brook; Harper Dewey, Sidney Center; A. L. White, North Walton; A. F. Getter, Masonville; Hon. I. E. Sherman, Sidney; Alton O. Potter, Franklin; C. H. Treadwell, Croton; E.C. Interlied, Rock Rift; Flavius Smith, Hancock

1889
President: C. B. Bassett
Vice President: Joseph Harby, Jr.
Secretary: The Rev. Scott B. Rathbun
Treasurer: J. H. Townsend
Directors: Alexander Tweedie, W. W. Ogden, R. Garrison, P. M. Hanford, W. C. Rogers, J. A. Robinson
Additional vice presidents were: George Lyons, Hobart; Zenas Mallory, Hamden; George Stillson, Franklin; Hon. Ira E. Sherman, Sidney; W. H. Benedict, North Walton; Edward C. Dean, Bovina; W. Miller, East Branch; E. C. Inderlied, Rock Rift; Hartson L. Wheat, Croton; William Tweedie, North Hamden; Flavius Smith, Hancock; Owen Priest, Masonville; Peter Dennis, Trout Creek; George D. Wheeler, Deposit; Harper Dewey, Sidney Centre; Alex Dysart, Delhi; A. C. Biggar, Colchester; C. Proskins, Fish's Eddy; Robert McGibbon, Cannonsville

1890
President: Joseph Harby, Jr.
Vice President: J. H. Townsend
Secretary: E. W. Pond
Treasurer and Purchasing Agent: J. W. St. John
Asst. Secy, Ladies' Dept.: Miss Mary D. St. John
Directors: R. Garrison, W. C. Rogers, W.W. Ogden, P. M. Hanford, J. A. Robinson, E. B. Howland, 2nd
Additional vice presidents were: George Lyon, Hobart; Zenas Mallory, Hamden; George Stillson, Franklin; Hon. Ira E. Sherman, Sidney; W. H. Benedict, North Walton; E. C. Inderlied, Rock Rift; Hartson L. Wheat, Croton; L. S. Mallory, East Branch; Thomas H. Winter, Margaretville; Parker Cook, Butternut Grove; Orin Priest, Masonville; S. N. Wheeler, Hancock; George D. Wheeler, Deposit; Harper Dewey, Sidney Centre; A. P. DuBois, Livingston Manor; Alexander Dysart, Delhi; A. C. Biggar, Colchester; C. Proskine, Fish's Eddy; Peter Dennis, Trout Creek; William Youmans, Delhi; William Tweedie, North Hamden; William Ruff, Bovina; H. S. Murray, Andes; Warner Sprague, Roscoe; A. S. Gilchrist, Meredith; Chandler Young, Liberty; Robert McGibbon, Cannonsville

1891
President: The Rev. Scott H. Rathbun
Vice President: J. H. Townsend
Secretary: E. W. Pond
Treasurer: W. A. Shepard
Directors: G. W. Wood, W. W. Ogden, John Holley, P. M. Hanford, J. A. Robinson, E. B. Howland, 2nd
Additional vice presidents were: Alexander Dysart, Delhi; Hartson L. Wheat, Croton; H. S. Murray, Andes; William Ruff, Bovina; George Lyon, Hobart; S. N. Wheeler, Hancock; A. C. Biggar, Colchester; Zenas Mallery, Hamden; Orrin Priest, Masonville; Warner Sprague, Roscoe; C. Proskine, Fish's Eddy; A. S. Gilchrist, Meredith; William Youmans, Delhi; Charles Potter, Franklin; William Porter, Unadilla; Chandler Young, Liberty; E. C. Inderlied, Rock Rift; Peter Dennis, Trout Creek; L. S. Mallory, East Branch; R. McGibbon, Cannonsville; George D. Wheeler, Deposit; Hon. Ira E. Sherman, Sidney; Harper Dewey, Sidney Centre; W. H. Benedict, North Walton; Parker Cook, Butternut Grove; William W. F. VanBenscoten, Margaretville

1895
President: A. F. Reynolds
Vice President: Archibald Tweedie
Secretary: Joseph Beckwith
Treasurer: W. A. Shepard
Directors: William W. Ogden, Lewis Bush, Ellis Taylor, John McGibbon, John Russell, George Webster
Additional vice presidents were: J. S. McMurdy, Delhi; M. J. Ward, Croton; E. J. Turnbull, Andes; Michael Miller, Bovina Center; George Lyon, Hobart; S. N. Wheeler, Hancock; A. C. Biggar, Colchester; W. W. Terry, Hamden; G. E. Smith;, Masonville; John Gibson, Fish's Eddy: John Mabee, Roscoe; E. C. Stuart, Franklin; W. W. Griffin, Trout Creek; Chandler Young, Liberty; Samuel Youmans, Rock Rift; William Potter, Unadilla; L. S. Mallory, East Branch; Alonzo Seymour, Cannonsville; H. B. Sewell, Sidney Center; W. H. Benedict, Northfield; C. A. Wheeler, Deposit; J. J. Watters, Sidney; Parker Cook, Butternut Grove; William Hendricks, Margaretville; Frank Johnson, Livingston Manor; A. M. Allen, Prattsville; E. E. Stebbins, Davenport Center; Amos Wood, North Hamden; Chauncey Alger, China; E. T. Finch, Granton; D. Anderson, Downsville; J. E. Miner, Shavertown

1896
President: A. F. Reynolds
Vice President: Archibald Tweedie
Secretary: Joseph Beckwith
Treasurer: W. A. Shepard
Directors: Harry K. Lewis, Lewis Bush, Ellis Taylor, John McGibbon, John Russell, George Webster
Additional vice presidents were: James Frasier, Frasier; M. J. Ward, Croton; D. Muir, Andes; Michael Miller, Bovina Center; George Lyon, Hobart; S. N. Wheeler, Hancock; A. C. Biggar, Colchester; W. W. Terry, Hamden; G. E. Smith, Masonville; George Smith, Fish's Eddy; John Gibson, Meredith; John Mabee, Roscoe; E. C. Stuart, Franklin; M. W. Griffin, Trout Creek; Chandler Young, Liberty; Samuel Youmans, Rock Rift; William Potter, Unadilla; L. S. Mallory, East Branch; Alonzo Seymour, Cannonsville; H.B. Sewell, Sidney Center; W. H. Benedict, Northfield; C. A. Wheeler, Deposit; J. J. Watters, Sidney; Parker Cook, Butternut Grove; William Hendricks, Margaretville; Frank Johnson, Livingston Manor; A. M. Allen, Prattsville; E. E. Stebbins, Davenport Center; Amos Wood, North Hamden; Chauncey Alger, China; E. T. Finch, Granton; N. O. Liddle, Downsville; J. E. Miner, Shavertown

1898
President: A. F. Reynolds
Vice President: C. A. Gould
Secretary: H. S. Sewell
Treasurer: W. A. Shepard
Directors: H. K. Lewis, Lewis Bush, E. B. Howland, John McGibbon, John Kilpatrick, Willard Dann
Additional vice presidents were: James Fraser, Fraser; M. J. Ward, Croton; George Cable, Andes; George Cable, Bovina Center; George A. Lyon, Hobart; S. N. Wheeler, Hancock; A. C. Biggar, Colchester; W. W. Terry, Hamden; G. E. Smith, Masonville; John Maybee, Roscoe; William Cole, Franklin; M. W. Griffin, Trout Creek; Hon. U. S. Messiter, Liberty; William Potter, Unadilla; L. S. Mallory, East Branch; Sherman Gregory, Cannonsville; H. B. Sewell, Sidney Centre; W. H. Benedict, Northfield; C. A. Wheeler, Deposit; J.J. Waters, Sidney; Parker Cook, Butternut Grove; Frank Johnson, Livingston Manor; A. M. Allen, Prattsville; E. E. Stebbins, Davenport Centre; George Jester, North Hamden; Delos Axtell, Barbourville; E. T. Finch, Granton; C. E. Hulbert, Downsville; J. E. Miner, Shavertown

1900
President: Wesley Ellis
Vice President: Robert Kilpatrick
Secretary: W. L. Gladstone
Treasurer: W. A. Shepard

Directors: W. R. Hall, E. B. Howland, John Kilpatrick, Willard Dann, Lewis Bush, John McGibbon
Additional vice presidents were: Peter Fraser, Fraser; M. J. Ward, Treadwell; James Norris, Shavertown; George Cable, Bovina Center; W. R. Brock, Hobart; George Beers, Sr., Hancock; Andrew E. Liddle, Andes; Ira Smith, Hamden; G. E. Smith, Masonville; John Maybee, Roscoe; Rufus Wood, Franklin; P. N. Dennis, Trout Creek; Hon. U.S. Messiter, Liberty; Joseph Hallenbeck, Unadilla; L. S. Mallory, East Branch; DeWitt Gregory, Cannonsville; H. W. Dewey, Sidney Centre; W. H. Benedict, Northfield; C. A. Wheeler, Deposit; J. J. Waters, Sidney; Parker Cook, Butternut Grove; Frank Johnson, Livingston Manor; A. M. Adien, Prattsville; E. E. Stebbins, Davenport Centre; John M. McGregor, North Hamden; Hon. Delos Axtell, Barbourville; Barton Miller, Granton; Thomas Holmes, Downsville; John B. Wyckoff, Grand Gorge

1902
President: A. F. Reynolds
Vice President: Robert Kilpatrick
Secretary: W. L. Gladstone
Treasurer: W. A. Shepard
Directors: W. R. Hall, E.B. Howland, W.D. Potter, C.A. Churchill, Lewis Bush, John McGibbon
Additional vice presidents were: Daniel Munn, Treadwell; John Shanley, Hancock; William Hymers, Hamden; G. E. Smith, Masonville; James Fitzgerald, Roscoe; Jacob Smith, Franklin; P. N. Dennis, Trout Creek; L. S. Mallory, East Branch; Frank R. Terry,Cannonsville; James W. Young, Sidney Centre; W. H. Benedict, Northfield; C. A. Wheeler, Deposit; J. J. Waters, Sidney; C. S. Wheat, North Hamden; Hon. Delos Axtel, Barbourville; Barton Miller, Granton; Thomas Holmes, Downsville

1903
President: A. F. Reynolds
Vice President: Robert Kilpatrick
Secretary: P. S. McNee
Treasurer: W.A. Shepard
Directors: W. R. Hall, E. R. Howland, W.D. Potter, C.A. Churchill, Lewis Bush, John McGibbon
Additional vice presidents were: Daniel Munn, Treadwell; John Shanley, Hancock; William Hymers, Hamden; G. E. Smith, Masonville; James Fitzgerald, Roscoe; Jacob Smith, Franklin; P. N. Dennis, Trout Creek; L. S. Mallory, East Branch; Frank R. Terry, Cannonsville; James W. Young, Sidney Centre; W. H. Benedict, Northfield; C. A. Wheeler, Deposit; J. J. Waters, Sidney; C. S. Wheat, North Hamden; Hon. Delos Axtell, Barbourville; Barton Miller, Granton; Thomas Holmes, Downsville

1909
President: A. F. Reynolds
Vice President: Robert Kilpatrick
Secretary: S. H. Osterhout
Treasurer: W. A. Shepard
Directors: C. A. Churchill, W. D. Potter John McGibbon, James A. Robinson, L. C. Bush, E. B. Howland
Additional vice presidents were: Daniel Munn, Walton; Clay V. More, Hancock; William Hymers, Hamden; G. E. Smith, Masonville; James Fitzgerald, Roscoe; Jacob Smith, Franklin; P. N. Dennis, Trout Creek; L. S. Mallory, East Branch; Stanton Gillette, Cannonsville; James W. Young, Sidney Centre; W. H. Benedict, Northfield; C. A. Wheeler, Deposit; J. J. Waters, Sidney; C. S. Wheat, North Hamden; Hon. Delos Axtell, Barbourville; Barton Miller, Granton; Thomas Tholmes, Downsville

1914
President: A. F. Reynolds
Vice President: V. A. Wolf
Secretary: E. D. Baker
Treasurer: S. H. Osterhout
Directors: John McGibbon, E. B. Howland, C. A. Churchill, W. D. Potter, William McDonald, Thomas Ford
Additional vice presidents were: Calvin J. Huson, Albany; George I. Wilbur, Oneonta; Clayton L. Wheeler, Hancock; E.A. Mackay, Franklin

1915
President: A. F. Reynolds
Vice President: V. A. Wolf
Secretary: E. D. Baker
Treasurer: S. H. Osterhout
Directors: William McDonald, Thomas Ford, H. D. Henderson, E. B. Howland, A. D. Scobie, W. D. Potter

1916
President: R. B. McFarlane
Vice President: W. H. Maynard
Secretary: A. L. Austin
Treasurer: Russell Archibald
Directors: Van E. Wilson, H. S. Hall, W. B. Thompson, J. J. Farrell, W. W. Hunt, C. E. Ladd
Additional vice presidents were: Charles S. Wilson, Albany; George I. Wilbur, Oneonta;

W. H. Hill, Binghamton; E. A. Mackay, Franklin

1923
President: W. H. Austin
Vice President: Frank Ives
Secretary: J. W. Palmer
Treasurer: Harry McCabe
Additional vice presidents were: Charles S. Wilson, Albany; Lincoln R. Long, New Kingston; W. H. Hill, Binghamton; A. J. Courtney, Walton

1932
President: J. J. Farrell
Vice President: Peter Chambers
Secretary: Harry McCabe
Treasurer: H. S. Ogden
Directors: E. F. Runnells, Harlan J. Wood, Fred R. Lyon, Louis F. Camp, C.S. Wheat, L. D. McClennon

1934
President: Peter Chambers
Vice President: Louis F. Camp
Secretary: Harry McCabe
Treasurer: H. S. Ogden
Directors: E.F. Runnells, Harlan J. Wood, J.J. Farrell, C.S. Wheat, L.D. McClennon

1935
President: Peter Chambers
Vice President: Louis F. Camp
Secretary: Harry McCabe
Treasurer: H. S. Ogden
Directors: E. F. Runnells, Harlan J. Wood, J. J. Farrell, C. S. Wheat, L. D. McClennon

1936
President: Peter Chambers
Vice President: D. W. Coulter
Secretary: C. O. Field
Treasurer: Louis Camp
Directors: Dr. E. Ogden Bush, Delhi; E. J. Kellam, Hancock; H. S. Ogden, Harry Main, Delhi; E. B. Guild

1937
President: Peter Chambers
Vice President: D. W. Coulter
Secretary: C. O. Field
Treasurer: Louis Camp
Directors: Dr. E. Ogden Bush, Delhi; E. J.Kellam, Hancock; Harry Main, Delhi; H. S. Ogden, E. B. Guild, Arthur Brundage, Margaretville

1939
President: Peter Chambers
Vice President: E. B. Guild
Secretary: Paul G. Williams
Treasurer: E. B. Guild
Directors: Dr. E. Ogden Bush, Delhi; E. J.Kellam, Hancock; Harry Main, Delhi; H. S. Ogden, C. O. Field, Arthur Brundage, Margaretville; William Webb

1944
President: Peter Chambers
Vice President: Harry Main, Delhi
Secretary: Paul G. Williams
Treasurer: E. B. Guild
Directors: E. J. Kellam, Hancock; H. S. Ogden, Louis Camp, C. O. Field, Arthur Brundage, Margaretville; Ralph S. Clark, Delhi

1945
President: Peter Chambers
Vice President: Harry Main, Delhi
Secretary: Paul G. Williams
Treasurer: E. B. Guild
Directors: E. J. Kellam, Hancock; H. S. Ogden, Louis Camp, C. O. Field, Arthur Brundage, Margaretville; Ralph S. Clark, Delhi

1946
President: Peter Chambers
Vice President: Harry Main, Delhi
Secretary: Paul G. Williams
Treasurer: E. B. Guild
Directors: E. J. Kellam, Hancock; H. S. Ogden, Louis Camp, C. O. Field, Arthur Brundage, Margaretville; Ralph S. Clark, Delhi

1947
President: Peter Chambers
Vice President: Harry Main, Delhi
Secretary: Paul G. Williams
Treasurer: E. B. Guild
Directors: E. J. Kellam, Hancock; H. S. Ogden, Louis Camp, C. O. Field, Arthur Brundage, Margaretville; Ralph S. Clark, Delhi

1948
President: Peter Chambers
Vice President: Harry Main
Secretary: Paul G. Williams
Treasurer: E. B. Guild
Directors: E. J. Kellam, Hancock; H. S. Ogden, Louis Camp, C. O. Field, Arthur Brundage, Margaretville; Ralph S. Clark, Delhi

1949
President: Peter Chambers
Vice President: Harry Main, Delhi
Secretary: Paul G. Williams
Treasurer: E. B. Guild
Directors: E. J. Kellam, Hancock; H. S. Ogden, Louis Camp, C. O. Field, Wallace Rich, Hobart; Russell Quackenbush, Sidney; Carlton Chambers

1950
President: Peter Chambers
Vice President: Harry Main, Delhi
Secretary: Paul G. Williams
Treasurer: E. B. Guild
Directors: E. J. Kellam, Hancock; H. S. Ogden, Louis Camp, C. O. Field, Wallace Rich, Hobart; Russell Quackenbush, Sidney; Carlton Chambers

1951
President: Peter Chambers
Vice President: Harry Main, Delhi
Secretary: Paul G. Williams
Treasurer: E. B. Guild
Directors: E. J. Kellam, Hancock; H. S. Ogden, Louis Camp, C. O. Field, Wallace Rich, Hobart; Russell Quackenbush, Sidney; Carlton Chambers

1952
President: Peter Chambers
Vice President: Wallace Rich, Hobart
Secretary: Paul G. Williams
Treasurer: E. B. Guild
Directors: William Bates, Sidney; H. S. Ogden, Louis Camp, C. O. Field, Carlton Chambers

1953
President: Peter Chambers
Vice President: Wallace Rich, Hobart
Secretary: Paul G. Williams
Treasurer: E. B. Guild
Directors: William Bates, Sidney; Ralph Higley, H. S. Ogden, Louis Camp, C. O. Field, Carlton Chambers

1954
President: Peter Chambers
Vice President: Wallace Rich, Hobart
Secretary: Paul G. Williams
Treasurer: E. B. Guild
Directors: William Bates, Sidney; Ralph Higley, H. S. Ogden, Louis Camp, C. O. Field, Carlton Chambers

1955
President: Peter Chambers
Vice President: Wallace Rich, Hobart
Secretary: Paul G. Williams
Treasurer: E. B. Guild
Directors: H. S. Ogden, Louis Camp, C. O. Field, Carlton Chambers, Ralph Higley

1956
President: Peter Chambers
Vice President: Wallace Rich, Hobart
Secretary: Paul G. Williams
Treasurer: E. B. Guild
Directors: H. S. Ogden, William Gifford, Deposit; Louis Camp, C. O. Field, Carlton Chambers, Ralph Higley

1957
President: Peter Chambers
Vice President: Wallace Rich, Hobart
Secretary: Paul G. Williams
Treasurer: E. B. Guild
Directors: H. S. Ogden, C. O. Field, Louis Camp, Ralph Higley, Carlton Chambers, William Gifford, Deposit

1958
President: George Rich, Franklin
Vice President: Graham Jackson
Secretary: John Elderkin
Treasurer: Milton Thomson
Directors: Carlton Chambers, Henry S. Ogden, Vanley Blackman, Herbert Elderkin, Steven Vamosy, Stamford; William Gifford, Deposit

1959
President: George Rich, Franklin
Vice President: Graham Jackson
Secretary: Ida Jackson Graby
Treasurer: Milton Thomson
Directors: Carlton Chambers, Vanley Blackman, Herbert Elderkin, Henry S. Ogden, Steven Vamosy, Stamford; William Gifford, Deposit

1960
President: George Rich, Franklin
Vice President: Graham Jackson
Secretary: Ida Jackson Graby
Treasurer: Milton Thomson
Directors: Vanley Blackman, Herbert Elderkin, Carlton Chambers,

William Tweedie, William Gifford, Deposit; Steven Vamosy, Stamford

1961
President: George Rich, Franklin
Vice President: Graham Jackson
Secretary: Ida Jackson Graby
Treasurer: Milton Thomson
Directors: Steven Vamosy, Stamford; Vanley Blackman, Herbert Elderkin, William Gifford, Deposit; Carlton Chambers, William Tweedie

1962
President: George Rich, Franklin
Vice President: Graham Jackson
Secretary: Ida Jackson Graby
Treasurer: Milton Thomson
Directors: Steven Vamosy, Stamford; Vanley Blackman, Herbert Elderkin,William Gifford, Deposit; Carlton Chambers, William Tweedie

1963
President: George Rich, Franklin
Vice President: Graham Jackson
Secretary: Ida Jackson Graby
Treasurer: Milton Thomson
Directors: Glenn Bowker, Steven Vamosy, Stamford; Vanley Blackman, Herbert Elderkin, William Gifford, Deposit; William Tweedie

1964
President: George Rich, Franklin
Vice President: Graham Jackson
Secretary: Ida Jackson Graby
Treasurer: Milton Thomson
Directors: Glenn Bowker, Steven Vamosy, Stamford; Vanley Blackman, Herbert Elderkin, William Gifford, Deposit; William Tweedie

1965
President: George Rich, Franklin
Vice President: Graham Jackson
Secretary: Ida Jackson Graby
Treasurer: Milton Thomson
Directors: Glenn Bowker, Steven Vamosy, Stamford; Vanley Blackman, Herbert Elderkin, William Gifford, Deposit; William Tweedie

1966
President: George Rich, Franklin
Vice President: Graham Jackson
Secretary: Ida Jackson Graby
Treasurer: Milton Thomson
Directors: Glenn Bowker, Steven Vamosy, Stamford; Vanley Blackman, Herbert Elderkin, William Gifford, Deposit; William Tweedie

1968
President: George Rich, Franklin
Vice President: Graham Jackson
Secretary: Ida Jackson Graby
Treasurer: Milton Thomson
Directors: Cecil Fowlston, Franklin; Glenn Bowker, Steven Vamosy, Stamford; Vanley Blackman, Herbert Elderkin, William Tweedie

1969
President: George Rich, Franklin
Vice President: Graham Jackson
Secretary: Ida Jackson Graby
Treasurer: Milton Thomson
Directors: Cecil Fowlston, Franklin, Glenn Bowker, Steven Vamosy, Stamford; Vanley Blackman, Herbert Elderkin, William Tweedie

1970
President: George Rich, Franklin
Vice President: Graham Jackson
Secretary: Ida Jackson Graby
Treasurer: Milton Thomson
Directors: Raymond Dann, Delhi; Cecil Fowlston, Franklin; Glenn Bowker, Vanley Blackman, Herbert Elderkin, William Tweedie

1971
President: George Rich, Franklin
Vice President: Graham Jackson
Secretary: Ida Jackson Graby
Treasurer: Milton Thomson
Directors: Raymond Dann, Delhi; Cecil Fowlston, Franklin; Glenn Bowker, Vanley Blackman,Herbert Elderkin, William Tweedie

1972
President: George Rich, Franklin
Vice President: Graham Jackson
Secretary: Ida Jackson Graby
Treasurer: Milton Thomson
Directors: Stanley Budine, Jr., Raymond Dann, Delhi; Cecil Fowlston, Franklin; Glenn Bowker,

Vanley Blackman, Herbert Elderkin

1973
President: George Rich, Franklin
Vice President: Graham Jackson
Secretary: Ida Jackson Graby
Treasurer: Milton Thomson
Directors: Stanley Budine, Jr., Raymond Dann, Delhi; Cecil Fowlston, Franklin; Glenn Bowker, Vanley Blackman, Herbert Elderkin

1974
President: George Rich, Franklin
Vice President: Graham Jackson
Secretary: Ida Jackson Graby
Treasurer: Milton Thomson
Directors: Stanley Budine, Jr., Raymond Dann, Delhi; Cecil Fowlston, Franklin; Glenn Bowker, Vanley Blackman, Herbert Elderkin

1975
President: George Rich, Franklin
Vice President: Graham Jackson
Secretary: Uva Scott
Treasurer: Milton Thomson
Directors: Stanley Budine, Jr., Raymond Dann, Delhi; Cecil Fowlston, Franklin; Glenn Bowker, Vanley Blackman, Herbert Elderkin

1976
President: George Rich, Franklin
Vice President: Graham Jackson
Secretary: Uva Scott
Treasurer: Milton Thomson
Directors: Stanley Budine, Jr., Raymond Dann, Delhi; Cecil Fowlston, Franklin; Glenn Bowker, Vanley Blackman, Herbert Elderkin

1977
President: George Rich, Franklin
Vice President: Graham Jackson
Secretary: Uva Scott
Treasurer: Milton Thomson
Directors: Norman Hoyt, Delhi; Stanley Budine, Jr., Cecil Fowlston, Franklin; Glenn Bowker, Vanley Blackman, Herbert Elderkin

1978
President: George Rich, Franklin
Vice President: Graham Jackson
Secretary: Vanley Blackman
Treasurer: Milton Thomson
Directors: Gary Tweedie, Norman Hoyt, Delhi; Stanley Budine, Jr., Cecil Fowlston, Franklin; Glenn Bowker, Herbert Elderkin

1979
President: George Rich, Franklin
Vice President: Norman Hoyt, Delhi
Secretary: Vanley Blackman
Treasurer: Milton Thomson
Directors: Graham Jackson, Gary Tweedie, Stanley Budine, Jr., Cecil Fowlston, Franklin; Glenn Bowker, Herbert Elderkin

1980
President: George Rich, Franklin
Vice President: Norman Hoyt, Delhi
Secretary: Arnold Froehlich
Treasurer: Milton Thomson
Directors: Arvetta McLaughlin, Robert McNee, DeLancey; Gary Tweedie, Stanley Budine, Jr., Glenn Bowker, Vanley Blackman

1981
President: George Rich, Franklin
Vice President: Gary Tweedie
2nd Vice President: Norman Hoyt, Delhi
Secretary: Arnold Froehlich
Treasurer: Milton Thomson
Directors: Doug Bull, DeLancey; Arvetta McLaughlin, Robert McNee, DeLancey; Vanley Blackman

1983
President: George Rich, Franklin
Vice President: Gary Tweedie
2nd Vice President: Norman Hoyt, Delhi
Secretary: Arnold Froehlich
Treasurer: Milton Thomson
Directors: Richard LaTourette, Sidney Center; Robert Cole, Delhi; Doug Bull, Arvetta McLaughlin, Robert McNee, DeLancey

1984
President: George Rich, Franklin
Vice President: Gary Tweedie
Secretary: Arnold Froehlich
Treasurer: Milton Thomson

Directors: Richard LaTourette, Sidney; Robert Cole, Delhi; Doug Bull, Arvetta McLaughlin, Robert McNee, DeLancey; Norman Hoyt, Delhi

1985
President: George Rich, Franklin
Vice President: Gary Tweedie
Secretary: Arnold Froehlich
Treasurer: Milton Thomson
Directors: Don Tweedie, James Taggart, Richard LaTourette, Sidney; Robert Cole, Delhi; Doug Bull, Arvetta McLaughlin

1986
President: George Rich, Franklin
Vice President: Don Tweedie
Secretary: Arnold Froehlich
Treasurer: Milton Thomson
Directors: John Jackson, Jon Seeley, Masonville; James Taggart, Richard LaTourette, Sidney; Robert Cole, Delhi; Doug Bull

1987
President: George Rich, Franklin
Vice President: Don Tweedie
Secretary: Arnold Froehlich
Treasurer: Milton Thomson
Directors: John Jackson, Jon Seeley, Masonville; James Taggart, Richard LaTourette, Sidney; Robert Cole, Delhi; Doug Bull

1988
President; George Rich, Franklin
Vice President: Don Tweedie
Secretary: Arnold Froehlich
Treasurer: Milton Thomson
Directors: Gary Tweedie, Max Dutcher, John Jackson, Jon Seeley, Masonville; James Taggart, Richard LaTourette, Sidney; Doug Bull

1989
President: George Rich, Franklin
Vice President: Don Tweedie
Secretary: Arnold Froehlich
Treasurer: Milton Thomson
Directors: Gary Tweedie, Max Dutcher, John Jackson, Jon Seeley, Masonville; James Taggart, Richard LaTourette, Sidney; Doug Bull

1990
President: George Rich, Franklin
Vice President: Don Tweedie
Secretary: Arnold Froehlich
Treasurer: Milton Thomson
Directors: Gary Tweedie, Max Dutcher, John Jackson, Jon Seeley, Masonville; James Taggart, Richard LaTourette, Sidney; Doug Bull

1991
President: Don Tweedie
Vice President: Gary Tweedie
Secretary: Arnold Froehlich
Treasurer: Milton Thomson
Directors: Niles Wilson, Max Dutcher, John Jackson, Jon Seeley, Masonville; James Taggart, Richard LaTourette, Sidney; Doug Bull

1992
President: Don Tweedie
Vice President: Gary Tweedie
Secretary: Arnold Froehlich
Treasurer: Milton Thomson
Directors: Roger Dibble, Hamden; Niles Wilson, Max Dutcher, John Jackson, Jon Seeley, Masonville; Richard LaTourette, Sidney; Doug Bull

1993
President: Don Tweedie
Vice President: Gary Tweedie
Secretary: Arnold Froehlich
Treasurer: Milton Thomson
Directors: Roger Dibble, Hamden; Niles Wilson, Max Dutcher, John Jackson, Jon Seeley, Masonville; Richard LaTourette, Sidney; Doug Bull

1994
President: Don Tweedie
Vice President: Gary Tweedie
Secretary: Gary Dreibelbis
Treasurer: Niles Wilson
Directors: Roger Dibble, Hamden; Max Dutcher, John Jackson, Jon Seeley, Masonville; Richard LaTourette, Sidney; Doug Bull, Arnold Froehlich

1995
President: Don Tweedie
Vice President: Richard LaTourette, Sidney
Secretary: Gary Dreibelbis
Treasurer: Niles Wilson
Directors: Roger Dibble, Hamden; Max Dutcher, John Jackson, Jon Seeley, Masonville; Doug Bull, Gary Tweedie, Arnold Froehlich

1996
President: Don Tweedie
Vice President: Richard LaTourette, Sidney
Secretary: Gary Dreibelbis
Treasurer: Niles Wilson
Directors: Roger Dibble, Hamden; Max Dutcher, John Jackson, Jon Seeley Masonville; Doug Bull, Gary Tweedie, Arnold Froehlich

1998
President: Don Tweedie
Vice President: Gary Tweedie
Secretary: Richard LaTourette, Sidney
Treasurer: Niles Wilson
Directors: Danny Hodges, L. Joseph Hoyt, Roger Dibble, Hamden; John Jackson, Jon Seeley, Masonville; Doug Bull

1999
President: Don Tweedie
Vice President: Gary Tweedie
Secretary: Richard LaTourette, Sidney
Treasurer: Niles Wilson
Directors: Bob Moody, Danny Hodges, L. Joseph Hoyt, Roger Dibble, Hamden; John Jackson, Jon Seeley, Masonville; Doug Bull

2000
President: John Jackson
Vice President: Richard LaTourette, Sidney
Secretary: L. Joseph Hoyt
Treasurer: Niles Wilson
Directors: Ed Rossley, Bovina Center; Pat Covell, Bob Moody, Danny Hodges, Roger Dibble, Hamden; Jon Seeley, Masonville; Doug Bull

2001
President: John Jackson
Vice President: Roger Dibble, Hamden
Secretary: L. Joseph Hoyt
Treasurer: Niles Wilson
Directors: Gary Tweedie, Ed Rossley, Bovina Center; Pat Covell, Bob Moody, Danny Hodges, Jon Seeley, Masonville; Doug Bull

Fair Boards Through the Years

Your Delaware County Fair Directors

PETER CHAMBERS
President

HARRY MAIN
Vice President

E. B. GUILD
Treasurer

PAUL G. WILLIAMS
Secretary

E. J. Kellam

Henry S. Ogden

C. O. Field

Louis F. Camp

Arthur Brundage

Ralph S. Clark

Walton Historical Society Photo

Fair directors in 1947.

Walton Historical Society Photo

The board of directors in 1889. Front row: The Rev. Scott B. Rathbun, secretary; Charles B. Bassett, the first president (he served from 1887 to 1889); board members James A. Robinson and William C. Rogers; Joseph Harby, Jr., vice president (president in 1895); and John H. Townsend, treasurer. Back row: Board members William W. Ogden, Platt M. Hanford and Richmond Garrison; J. Barton Miller, who served in the horse department in 1889 and became general superintendent of the Fair in 1891; and board member Alexander Tweedie.

Walton Historical Society Photo

In 1961, pictured, front row, George Rich, president; Ida Jackson Graby, secretary and Milton Thomson, treasurer. Second row, Vanley Blackman, Graham Jackson, vice president; Herbert Elderkin, William Tweedie and Steven Vamosy.
Not pictured, Carlton Chambers and William Gifford.

Walton Historical Society Photo

These were the Fair directors in 1979. The officers: Milton Thomson, treasurer; George Rich, president; Norman Hoyt, vice president; and Vanley Blackman, secretary. Back row, Cecil Fowlston, Glenn Bowker, Stanley Budine, Jr., Gary Tweedie and Graham Jackson. Not pictured, Herbert Elderkin.

In the Centennial Year of 1986, the board was composed of, front row, Donald Tweedie, vice president; George Rich, president; and Milton Thomson, treasurer. Back row, Arnold Froehlich, secretary; Douglas Bull, Gary Tweedie, James Taggart, Robert Cole and Jon Seeley. Not pictured Richard LaTourette and John Jackson.

Notes

Part I: The Fair Years

The First Fairs

1 *History of Delaware County, N.Y. with Illustrations, Biographical Sketches and Portraits of Some Pioneers and Prominent Residences.* New York: W. W. Munsell and Co., 1880 p. 324.

2 DeNatale, Douglas. *Two Stones for Every Dirt.* Fleischmanns, N.Y.: Purple Mountain Press, Ltd. 1987 p. 24.

3 Lane, Frank and Helen. *The Story of Walton: 1785 to 1975.* Walton, N.Y.: The Walton Historical Society, 1975, p. 72.

4 *Ibid.*, pp 26 - 29.

5 *Two Stones for Every Dirt*, p. 49.

6 Ida Graby in *The Story of Walton*, p. 185.

7 Munsell, p. 156.

The Fair in Walton

1 *The Story of Walton*, p. 29.

2 *Ibid.*, p. 185.

3 *The Walton Reporter*, Walton, N.Y., October 13, 1887.

4 *Ibid.*, August 18, 1961.

The Years Fly By

1 *The Walton Reporter*, September 12, 1931, P. 1.

2 *The Daily Star* (Oneonta, N.Y.), August 2, 1999, P. 1.

3 *The Walton Reporter*, August 25, 1939, P. 1.

4 *75 Years of Fun & Friendship*, James Rice, ed., Cornell Cooperative Extension, 1993.

1957: A Time of Crisis

1 *The Daily Star* (Oneonta), May 5, 1957.

2 *The Walton Reporter*, February 18, 1957.

3 *Ibid.*, December 20, 1957.

Reaching the Millennium

1 *The Daily Star* (Oneonta), August 12, 1999, p. 12.

Youth: The Future of the Fair

1 *Transactions of the New York State Agricultural Society*, Vol. VII. 1847. Albany, N.Y.: G. Van Benthuysen, public printer. 1848, p. 17.

2 *75 Years of Fun & Friendship*, p. 94.

3 *The Walton Reporter*, August 27, 1943.

Part II: Under the Tents and Inside the Track

Dairy Winners and Quality Beef

1 Purdy, Herman R. *Breeds of Cattle*. Completed by R. John Dawes. New York: Chanticleer Press, Inc. 1987, p. 164.

2 *Ibid.*, p. 178.
Also, U.S. Department of Agriculture Farmers' Bulletin No. 2228 *Beef Cattle Breeds*, p 8.

3 *Breeds of Cattle*. p. 213.

4 *Ibid.*, p. 237-239.

5 *Ibid.*, p. 83.

6 *Ibid.*, p, 225, 229.

7 *Ibid.*, p. 247.

8 *Ibid.*, p. 71, 72.

9 *Ibid.*, p. 22.
Also, Mansfield, Richard H. *Progress of the Breed*. Centennial Edition. Robert H. Hastings, ed. Sandy Creek, N.Y.: Holstein-Friesian World, Inc. 1985 p. 8-10.

10 Prescott, Maurice S., ed. *Holstein-Friesian History*. Diamond Jubilee Edition. Lacona, N.Y: Holstein-Friesian World, Inc 1960 p. 9, 10, 54
Also, *Progress of the Breed*, p. 8, 16.

11 *Breeds of Cattle*. P. 220.

12 *Progress of the Breed*. P. 12.

13 *Ibid.*, p. 45.

14 *Ibid.*, p. 89.

15 *Ibid.*, p. 245.

The Farm Trot and On From There

1 "The Chinese Shanghai fowl came to England and America in 1845. The name of this Asiatic breed was later changed to Cochin." *American Standard of Perfection.* Mendon, Mass., American Poultry Association, Inc. 1998 edition.

2 Howard, Tom. *The Illustrated Horse.* Edison, N.J.: Chartwell Books, Inc. A Division of Book Sales, Inc. 1994 p. 62.

3 Braider, Donald. *The Life, History and Magic of the Horse.* New York: Grosset & Dunlap. 1973 p. 134.

4 Glyn, Richard. *The World's Finest Horses and Ponies.* Garden City, N.Y.: Doubleday & Co., 1971 p. 77.

5 *The Illustrated Horse,* p. 6.

6 *The Walton Reporter.* 1933.

Those Ancient and Lovable Goats

1 Drummond, Susan Black. *Angors Goats thee Northern Way.* Freeport, Mich.: Stony Lonesome Farm. Fourth Edition. 1993 p. 13.

2 Belanger, Jerry, *Raising Milk Goats the Modern Way.* Pownal, Vermont: Storey Communications, Inc. Revised Edition. 1990 p. 9.

3 *Raising Milk Goats the Modern Way.* p. 6.

4 *Ibid.,* p. 9.

5 *Ibid.,* p. 5.

6 *Ibid.,* p. 7.

7 *Ibid.,* p. 9.

Chickens Are Not What They Used To Be

1 *The Poultry Tribune.* Mt. Morris, Ill.: Watt Publishing Co. September, 1995 p. 2.

2 *Ibid.,* p. 2.

3 *Ibid.,* p. 2.

4 *American Standard of Perfection.* P. 41.

Bibliography

Adams, Arthur G. *The Catskills - A Guide to Mountains and Nearby Villages.* Fleischmanns, N.Y.: Purple Mountain Press, 1988.

American Livestock Breeds Conservancy, P.O. Box 477, Pittsboro, North Carolina 27312.

Beef Cattle Breeds, United States Department of Agriculture Farmers' Bulletin No. 2228. Washington, D.C.: U.S. Govt. Printing Office, Revised 1975.

Belanger, Jerry. *Raising Milk Goats the Modern Way.* Revised ed. Pownal, Vt.: Storey Communications, Inc., 1992.

Braider, Donald. *The Life, History and Magic of the Horse.* New York: Madison Square Press, Grosset & Dunlap, 1973.

Columbia Encyclopedia, 2nd Edition. New York: Columbia University Press. 12th printing, 1956.

Delaware County Fair Souvenir Program Books. Delhi, N.Y.: Meredith Productions, 1993, 1994.

Delaware County Times. Delhi, N.Y.

Delaware Gazette. Delhi, N.Y.

Delaware Republican. Delhi, N.Y.: July 3, 1869, "Fair Held in Walton."

Delaware Valley Agricultural Society Premium Books: 1887, 1888, 1890, 1891. 1903, 1906, 1909, 1914, 1916, 1923, 1932, 1934, 1936, 1948, 1955, 1957, 1958 1960 - 1969, 1974 - 1979, 1980, 1981, 1983, 1985 - 1989, 1995, 1996, 1998, 1999, 2000.

DeNatale, Douglas. *Two Stones for Every Dirt - The Story of Delaware County, N.Y.* Delaware County Historical Association. Fleischmanns, N.Y.: Purple Mountain Press, 1987.

Drummond, Susan Black. *Angora Goats the Northern Way.* Freeport, Mich.: Stony Lonesome Farm, 4th ed., 1993.

Francis, Marilyn Milow, "See You at the Fair." *Kaatskill Life (*magazine) Delhi, N.Y.: Summer, 1996, pp 17-22.

Glyn, Richard. *The World's Finest Horses and Ponies.* Garden City, N.Y.: Doubleday & Co., 1971.

Gould, Jay. *History of Delaware County.* Roxbury, N.Y.: Keeny and Gould, 1856.

Graby, Ida. *History of Delaware County Fairs.* Walton, N.Y.: Ida Graby, 1975.

History of Delaware County, N.Y. with Illustrations, Biographical Sketches and Portraits of Some Pioneers and Prominent Residences. New York: W. W. Munsell & Co., 1880.

Holstein World (magazine). Sandy Creek, N.Y.: Dairy Business Communications, March, 2000.

Horton, Gertrude Fitch. *Old Delaware County: A Memoir.* Delhi, N.Y.: Delaware County Historical Association.

Howard, Tom. *The Illustrated Horse.* Edison, N.J.: Chartwell Books, Inc., 1994.

Lane, Frank and Helen. *Walton Yesteryears.* Walton, N.Y.: Walton Historical Society, The Reporter Co., Inc., 1985 (Reprinted in 1990).

Lennox, Florence S. and Warner, Robert. "Delaware Has Had Fair for More Than Century," *The Oneonta Star,* Oneonta, N.Y.: August 18, 1952.

Mackenzie, David. *Goat Husbandry.* Norfolk, England: Fakenham Press Ltd., 4th ed., 1980.
Mansfield, Richard H. *Progress of the Breed: The History of U.S. Holsteins.* Edited by Robert H. Hastings. Sandy Creek, N.Y.: Holstein-Friesian World, Inc., Centennial Edition, 1985.
Monroe, John D. *Chapters in the History of Delaware County. New York.* Delhi, N.Y.: Delaware County Historical Association, 1949.
Nash, Marion R. *The Stringman's Scrap Book.* 2nd edition, 1979.
New York State Agricultural Society. *Transactions,* Vol. VII. Albany, N.Y.: C. Van Benthuysen, 1848.
Nicholds, Elizabeth. *Thunder Hill.* Garden City, N.Y.: Doubleday & Co., Inc., 1953.
—— "The Goats of Thunder Hill." *The National Geographic Magazine.* May, 1954, pp. 625-640.

Prescott, Maurice S. and Scholl, Melvin. *Holstein-Friesian History.* Lacona, N.Y.: Holstein-Friesian World, Inc., Diamond Jubilee Edition, 1960.
Purdy, Herman R. and Dawes, R. John. *Breeds of Cattle.* New York: Chanticleer Press, Inc., 1987

Raitt, John E. *Ruts in the Road.* Delhi, N.Y.: John E. Raitt, 1982.
___ *The Old Watering Trough.* Delhi, N.Y.: 1982.
Rice, James, ed. Cornell Cooperative Extension. *75 Years of Fun & Friendship.* Walton, N.Y.: The Reporter Co., 1993.

The Daily Star, Oneonta, N.Y.: *100th Year Supplement.* August 6, 1986.
The Delaware Republican. Delhi, N. Y.: "List of Premiums for 1869." July 3, 1869.
The Poultry Tribune. Mt. Morris, Illinois: Watt Publishing Co., Sept. 1995, p. 2.
The Walton Reporter. Walton, N.Y.: 100th Year Supplement, August 6, 1986.
The Walton Blade. Walton, N.Y.: "What Is the Object of an Agricultural Fair?" October 7, 1856.
———— "The County Fair": List of Premium Awards. September 30, 1856.
The Walton Chronicle. Walton, N.Y.: June 16, July 21, August 11 and 18, October 8, 1887.
The Walton Reporter. Walton, N.Y.: miscellaneous clippings, 1887-2001.

Wakeman, Richard P. *A History of the Delaware Valley Agricultural Society as it Celebrates its 100th Fair, July 1887 to August 1986.* Walton, N.Y.: The Reporter Co., 1986.
Walton Historical Society. *The Story of Walton, 1785-1975.* Deposit, N.Y.: Valley Offset, Inc., 1975. Second Printing, Deposit, N. Y.: Courier Printing Corp., 1997.